ELECTRONIC INVENTIONS
1745-1976

Other Titles of Interest

ELECTRONIC INVENTIONS
1745-1976

by

G. W. A. DUMMER
M.B.E.,C.Eng.,F.I.E.E.,F.I.E.E.E.,F.I.E.R.E.
(formerly Supt. Applied Physics, Royal Signals & Radar Establishment)

PERGAMON PRESS
OXFORD · NEW YORK · TORONTO · SYDNEY · PARIS · FRANKFURT

U.K.	Pergamon Press Ltd., Headington Hill Hall, Oxford OX3 0BW, England
U.S.A.	Pergamon Press Inc., Maxwell House, Fairview Park, Elmsford, New York 10523, U.S.A.
CANADA	Pergamon of Canada Ltd., 75 The East Mall, Toronto, Ontario, Canada
AUSTRALIA	Pergamon Press (Aust.) Pty. Ltd., 19a Boundary Street, Rushcutters Bay, N.S.W. 2011, Australia
FRANCE	Pergamon Press SARL, 24 rue des Ecoles, 75240 Paris, Cedex 05, France
WEST GERMANY	Pergamon Press GmbH, 6242 Kronberg-Taunus, Pferdstrasse 1, Frankfurt-am-Main, West Germany

First edition 1977

Library of Congress Cataloging in Publication Data

Dummer, Geoffrey William Arnold.
Electronic inventions from 1745-1976.

Includes indexes.
1. Electronics--History. 2. Electronic apparatus
and appliances--History. I. Title.
TK7809.D85 1976   621.381'09   76-15157
ISBN 0-08-020982-3

In order to make this volume available as economically and rapidly as possible the author's typescript has been reproduced in its original form. This method unfortunately has its typographical limitations but it is hoped that they in no way distract the reader.

Printed in Great Britain by Biddles Ltd., Guildford, Surrey

Contents

Preface

In 1970 the author published a Survey Paper entitled "Electronic Components - Past, Present and Future - from 1745 to the year 2000", which summarised the author's fifty years experience in component development, reliability of electronics and introduction of microelectronics. Now in semi-retirement, it was thought useful to continue this work and to collect and collate the dates of inventions over the whole field of electronics. Requests for some of these dates have arisen in the case of patent litigations and others are of interest to research organisations, Universities and manufacturers as a source of reference. It is not intended that this book should be a learned treatise on a particular aspect of historial electronics, such as those written by Prof. Susskind in America and Prof. Tucker in the U.K. but rather a summary of first dates in electronics development over a very wide field for ready reference.

Because no one person can be an authority in all fields of electronics, the data given is extracted from a wide variety of published sources, i.e. books, patents, technical journals, proceedings of Societies, etc., to whom full acknowledgement is made. This present work covers inventions from Europe and the U.S.A. and does not include Japan and Russia. Obviously a survey such as this cannot be completely accurate because of - in some cases - the passage of time, and in others, conflicting claims. The author would, therefore, be glad to add additional data, provided the published source is quoted, in a further edition.

How does one define an electronic invention Bearing in mind the average electronic engineer's interest in his own particular field, the first person to initiate or develop a new technique concerned with electronics has been included in this edition and the selection made on the basis of simple language and explanation.

The history of electronics is a fascinating story of developments changing from electricity into electronics with no readily definable dividing line.

The early 18th and 19th century inventors concerned with electricity who laid the foundations of electronics had inquiring minds and - as the pace of life was then slower - had usually time to develop their ideas. Men like Faraday, Rutherford, Thomson, Maxwell and others were products of their time in the history of invention and were the inventive giants of their day.

Later came the theoretical physicists - Planck, Bohr, Lorentz, Einstein and many others who quantified the work of their predecessors and delved into the reasons why, rather than achieving practical results.

Starting around the 1920's, a host of inventors developed electronic ideas and produced devices which are still in use today - iconoscope, klystron, magnetron, electron microscope, etc., By the 1950's the swell developed into a torrent, with the invention of the transistor revolutionising valve oriented electronics. Today the pace of development is still increasing - satellites, computers, television, optoelectronics and the number and scope of inventions is still growing.

The wealth created by electronic inventions is very large indeed and has gone to the country which has capitalised on its inventors. The effect on our daily lives of telephones, television and computers has not only created wealth, but given job opportunities to many. The process of invention has changed from the individual inventor to that of the large research laboratories which have the advantages of funds and cross-fertilisation of ideas. Certainly the Bell Laboratories team in the U.S.A. made the greatest contributions to semiconductor

1

technology, not only by inventing a working transistor, but by producing materials (Si, Ge) of a purity previously unknown. This work, basic to microelectronics, has created entirely new industries. The complexity of modern electronics has also brought together chemists, physicists, mathematicians, engineers and others as the fields of development widen. Research, development and production are now more closely integrated.

The place of the inventor in modern electronics seems to be in the large laboratory, but there are always exceptions to the rule. The 295 inventions described in this book, together with almost 500 additional references, form a background to electronics progress which, in ever increasing tempo, is now changing the world we live in.

G. W. A. Dummer.

27 King Edwards Road, Malvern Wells, Worcs. WR14 4AJ, England.

Acknowledgements

My task in this book has been that of a compiler rather than that of an author and this has only been made possible because of the co-operation of so many authorities. In particular the IEEE has been most helpful, as the 50th Anniversary edition of the Proc. IRE provided a mine of information on early electronics. Thanks are also due to the IERE and the IEE for permission to quote extracts from their publications.

Many books and technical journals have provided extracts which are relevant and the author is indebted to all those detailed in the "SOURCE" following each extract. Where "SOURCE" is quoted the words and opinions are exactly those of the authors of the extracts. The page number given in each case is that of the extract and not that of the title page. Full acknowledgement is made to all authors quoted. Thanks are due to many books and journals for their permission to quote from their publications and also to the Patent Office and to Libraries for their help.

The author would like to record his appreciation of the help given by the Science Museum London, in particular Mr. W. K. E. Geddes, Dr. Denys Vaughan and Dr. B. P. Bowers and also the following for their advice and assistance: S. W. Amos, W. Bardsley, P. T. Baxendall, C. den Brinker, T. A. Everist, C. Hilsum, T. P. McLean and E. H. Putley.

It is hoped that the data patiently collected for this book will be found useful, both as a source of reference and as a review of electronics development.

Chapter 1.   Charts of Electronic inventions

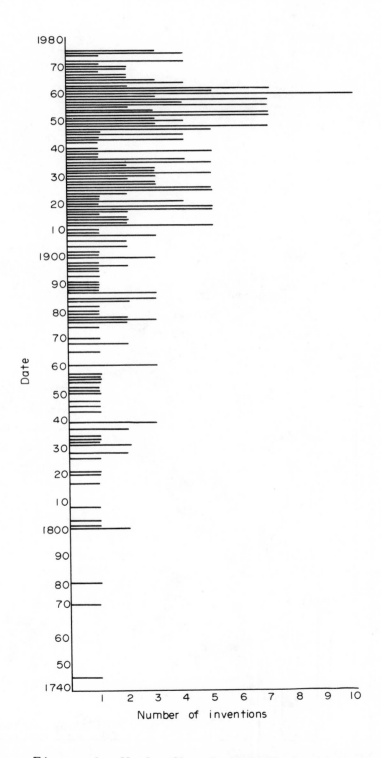

Figure 1   Number of inventions plotted by date from 1745 to 1976.

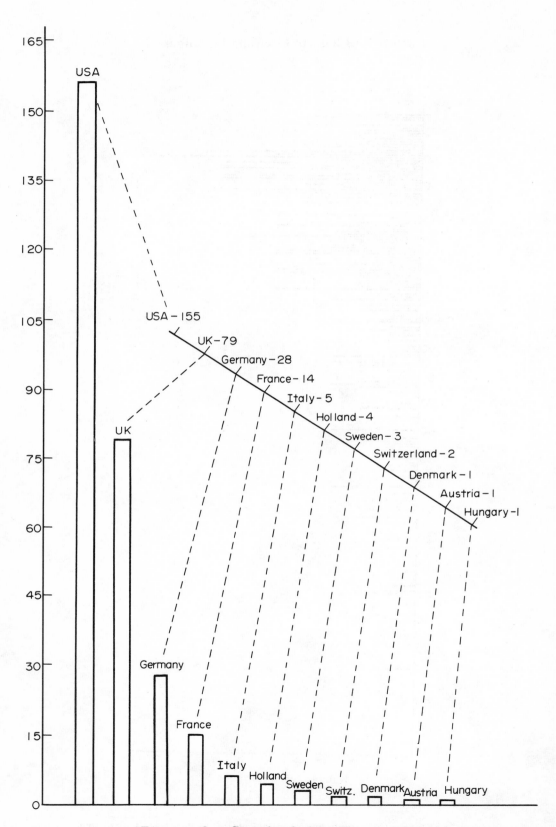

Figure 2 Countries of origin of inventions (Europe and U.S.A.)

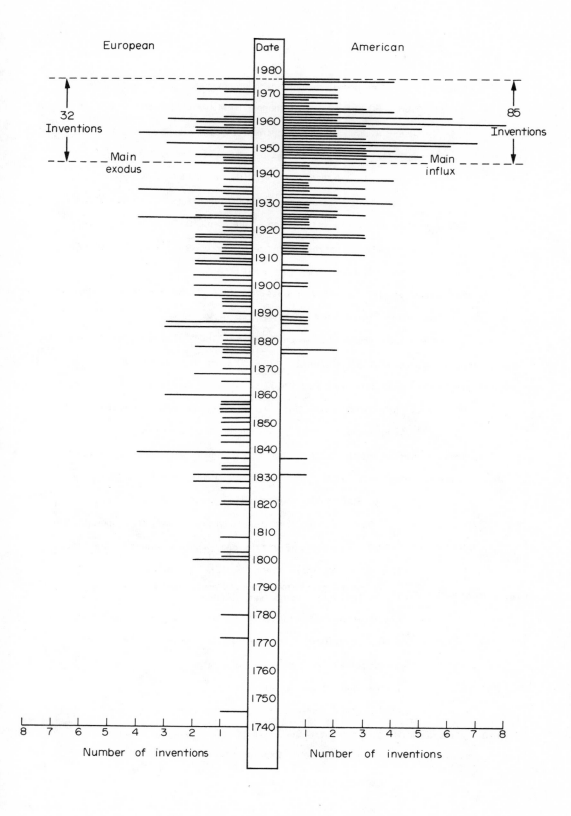

Figure 3   **The effect of the "brain drain" from Europe to U.S.A.**

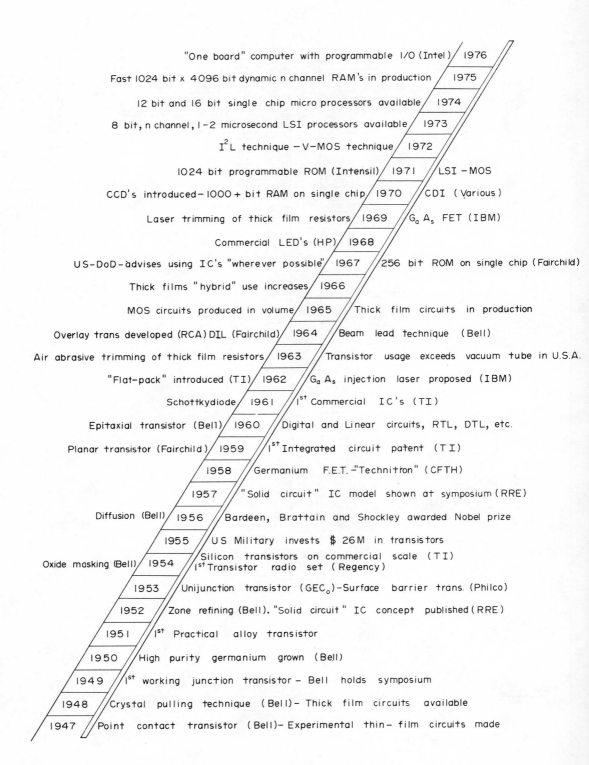

"One board" computer with programmable I/O (Intel) / 1976

Fast 1024 bit x 4096 bit dynamic n channel RAM's in production / 1975

12 bit and 16 bit single chip micro processors available / 1974

8 bit, n channel, 1–2 microsecond LSI processors available / 1973

I^2L technique — V–MOS technique / 1972

1024 bit programmable ROM (Intensil) / 1971 / LSI – MOS

CCD's introduced – 1000 + bit RAM on single chip / 1970 / CDI (Various)

Laser trimming of thick film resistors / 1969 / $G_a A_s$ FET (IBM)

Commercial LED's (HP) / 1968

US–DoD–advises using IC's "wherever possible" / 1967 / 256 bit ROM on single chip (Fairchild)

Thick films "hybrid" use increases / 1966

MOS circuits produced in volume / 1965 / Thick film circuits in production

Overlay trans developed (RCA) DIL (Fairchild) / 1964 / Beam lead technique (Bell)

Air abrasive trimming of thick film resistors / 1963 / Transistor usage exceeds vacuum tube in U.S.A.

"Flat-pack" introduced (TI) / 1962 / $G_a A_s$ injection laser proposed (IBM)

Schottkydiode / 1961 / 1st Commercial IC's (TI)

Epitaxial transistor (Bell) / 1960 / Digital and Linear circuits, RTL, DTL, etc.

Planar transistor (Fairchild) / 1959 / 1st Integrated circuit patent (TI)

1958 / Germanium F.E.T. –"Technitron" (CFTH)

1957 / "Solid circuit" IC model shown at symposium (RRE)

Diffusion (Bell) / 1956 / Bardeen, Brattain and Shockley awarded Nobel prize

1955 / US Military invests $ 26M in transistors

Oxide masking (Bell) / 1954 / Silicon transistors on commercial scale (TI)
1st Transistor radio set (Regency)

1953 / Unijunction transistor (GEC_o)–Surface barrier trans. (Philco)

1952 / Zone refining (Bell). "Solid circuit" IC concept published (RRE)

1951 / 1st Practical alloy transistor

1950 / High purity germanium grown (Bell)

1949 / 1st working junction transistor – Bell holds symposium

1948 / Crystal pulling technique (Bell)– Thick film circuits available

1947 / Point contact transistor (Bell)– Experimental thin–film circuits made

Figure 4   **The ladder of microelectronics development.**

In the late 1960s, the following chart was produced
showing packing density achieved with components available
from 1940 to that date. This early chart, although still in
parts per cubic inch, has been brought up to date as far as
is possible. Although LSI, CCD's and bubble devices are
not normally stacked in one cubic inch, the chart illustrates
the fantastic reduction in size which has taken place over
the last few years.

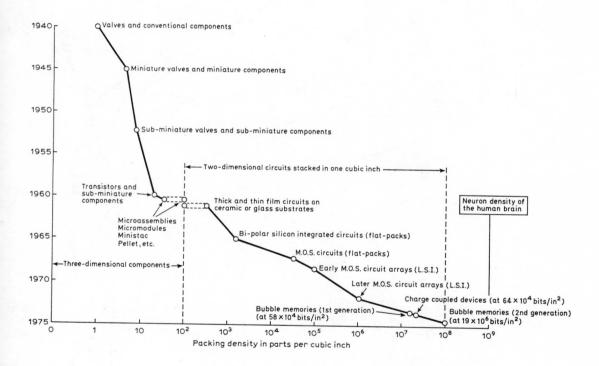

Figure 5   Electronics miniaturisation in terms of packing densities.

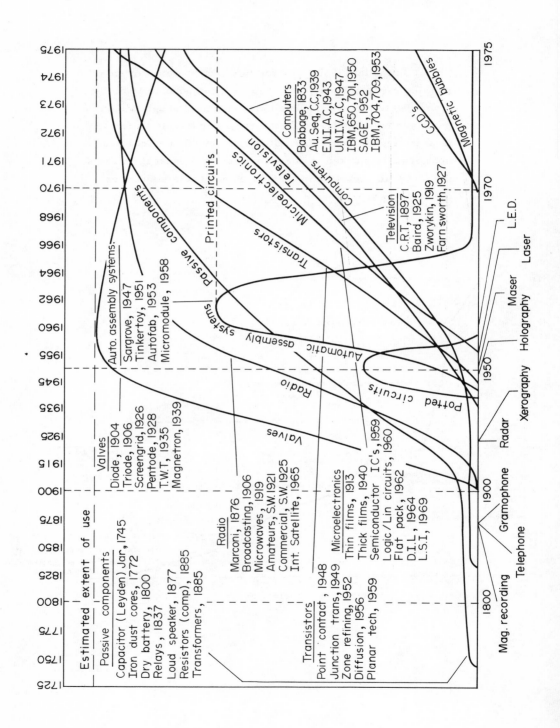

Figure 6   The rise and fall of electronics technologies.

Chapter 2. List of inventions in date order

N.B. The date can be used as an approximate index, as Chapter 5 is arranged in chronological order.

1642	Calculating machine	(Pascal)
1672	Calculating machine	(Leibniz)
1745	Capacitor	(Leyden Jar)
1772	Iron dust cores	(Knight)
1780	Galvanic action	(Galvani)
1800	Dry battery	(Volta)
1800	Infra-red radiation	(Herschel)
1801	Ultra-violet radiation	(Ritter)
1803	Accumulator	(Ritter)
1808	Atomic theory	(Dalton)
1820	Electro-magnetism	(Oersted)
1821	Thermoelectricity	(Seebeck)
1826	Ohms law	(Ohm)
1828	Moving coil galvanometer	(Schweigger)
1828	Astatic galvanometer	(Nobilli)
1831	Electromagnetic induction	(Faraday)
1831	Transformer	(Faraday)
1832	Self-induction	(Henry)
1833	Calculating machine	(Babbage)
1834	Electrolysis	(Faraday)
1837	Relays	(Cooke, Wheatstone, Davy)
1837	Electric motor	(Davenport)
1839	Magnetohydrodynamic generation	(Faraday)
1839	Fuel cell	(Grove)
1839	Microfilming	(Dancer)
1839	Photovoltaic effect	(Becquerel)
1843	Fascimile reproduction	(Bain)
1845	Kirchoff's laws	(Kirchoff)
1847	Magnetostriction	(Joule)
1850	Thermistor	(Faraday)
1851	Magnetism electricity theory	(Kelvin)
1852	Sputtering process	(Grove)
1854	Calculating machine	(Scheutz)
1855	Cold cathode discharge tube	(Gaugain)
1856	Low pressure discharge tube	(Geissler)
1857	Mercury arc lamp	(Way)
1860	Secondary battery - accumulator	(Planté)
1860	Diaphragm microphone	(Reis)
1860	Electron	(Thomson)
1865	Radio wave propagation	(Maxwell)
1868	Battery, Léclanché cell	(Leclanche)
1868	Rheotome	(Lenz)

1870	Battery, Clark standard cell	(Clark)
1874	Mica capacitor	(Bauer)
1876	Rolled paper capacitor	(Fitzgerald)
1876	Telephone	(Bell)
1877	Phonograph	(Edison)
1877	Carbon microphone	(Edison)
1877	Moving coil loudspeaker	(Siemens)
1878	Cathode rays	(Crookes)
1878	Carbon filament incandescent lamp	(Swan, Stearn, Topham & Cross)
1879	Hall effect	(Hall)
1880	Piezo electricity	(Curies)
1882	Wimshurst machine	(Wimshurst)
1884	Zinc-mercuric-oxide battery	(Clarke)
1884	Quick-break switch	(Holmes)
1885	Moulded carbon composition resistor	(Bradley)
1885	Distribution transformer	(Deri)
1885	Power transformer	(Zipernowski, Deri & Blathy)
1887	Gramophone	(Berliner)
1887	Aerial	(Hertz)
1887	Quick make and break switch	(Holmes)
1888	Induction motor	(Tesla)
1889	Tabulating machine	(Hollerith)
1890	Coherer	(Branley)
1891	Battery, Weston standard cell	(Weston)
1893	Waveguide theory	(Thomson)
1895	X-rays	(Rontgen)
1896	Wireless telegraphy	(Marconi)
1897	Carbon film resistor	(Gambrill)
1897	Cathode ray oscillograph	(Braun)
1898	Magnetic recording	(Poulsen)
1900	Battery, nickel-iron cell	(Edison)
1900	Battery, nickel-cadmium cell	(Junger & Berg)
1900	Ceramic capacitor	(Lombardi)
1901	Fluorescent lamp	(Cooper-Hewitt)
1902	Synchronous induction motor	(Danielson)
1904	Tubular glass capacitor	(Mosciki)
1904	Two electrode valve	(Fleming)
1906	Radio broadcasting	(Fessenden)
1906	Three electrode valve	(de Forest)
1908	Geiger counter	(Rutherford, Geiger)
1908	Electric organ	(Cahill)
1908	Television, electronic theory	(Campbell-Swinton)
1909	Ferrites	(Hilpert, Snoek)
1910	Neon lamp	(Claude)
1912	Regenerative circuit	(de Forest, Armstrong, Langmuir)
1912	Tungar rectifier	(Langmuir)
1912	Radio-atmospheric propagation	(Eccles)
1912	Heterodyne and superheterodyne circuit	(Fessenden, Armstrong)

Year	Invention	Inventor
1912	Cloud chamber	(Wilson)
1913	Reliability standards	(AIEE)
1913	Metal-film resistor	(Swann)
1914	Ultrasonics	(Langevin)
1914	Thyratron	(Langmuir)
1915	Electromagnetic filter	(Campbell, Wagner)
1915	Radio - single sideband communication	(Carson)
1916	Reliability control	(Bell/Western Electric)
1917	Condenser microphone	(Wente)
1917	Crystal pulling technique	(Czochralski)
1918	Neutrodyne circuit	(Hazeltine)
1918	Radio, ground wave propagation	(Watson)
1919	Multivibrator circuit	(Abraham & Bloch)
1918	Shot effect noise	(Schottky)
1918	Induction heating	(Northrup)
1919	Spiralled metal film resistor	(Kruger)
1919	Flip-flop circuit	(Eccles & Jordan)
1919	Microwave oscillator	(Barkhausen & Kurtz)
1919	Valves, houskeeper seal	(Houskeeper)
1919	Television, electronic system	(Zworykin, Farnsworth)
1920	Ultra-micrometer	(Whiddington)
1921	Amateur short-wave radio	(Various)
1921	Ferroelectricity	(Valasek)
1921	Crystal control of frequency	(Cady)
1922	Negative resistance oscillator	(Gill, Morrell)
1923	Iconoscope	(Zworykin)
1924	Saw tooth time base circuit	(Anson)
1924	Radar	(Appleton, Briet, Watson-Watt et al)
1925	Electrostatic loudspeaker	(Various)
1925	Cracked-carbon resistor	(Siemens & Halske)
1925	Johnson noise	(Johnson)
1925	Commercial short-wave radio	(Philips)
1925	Television, mechanical scanning	(Baird)
1926	Film sound recording - sound-on-disc	(Warner Bros.)
1926	Automatic volume control circuit	(Wheeler)
1926	Yagi aerial	(Yagi)
1926	Sprayed metal film resistor	(Loewe)
1926	Screened grid valve	(Round)
1927	Negative feedback amplifer	(Black)
1927	Film sound recording - sound-on-film	(Fox Movietone News)
1927	Television, electronic system (see 1919)	(Farnsworth - see 1919)
1928	Diversity reception	(Beverage, Peterson, Moore)
1928	Quartz clock frequency standard	(Horton, Morrison)
1928	Pentode valve	(Tellegen, Holst)
1929	Microwave communication	(Clavier)
1929	Cyclotron	(Laurence)
1930	Van de Graaf accelerator	(Van de Graaf)
1930	Transistor, MOS/FET	(Lilienfeld)

1930s	Radiophonic sound and music	(Grainger)
1931	Oxide film resistor	(Littleton)
1931	Stereophonic sound	(Blumlein, Bell Labs.)
1931	Computer - differential analyser	(Bush)
1931	Reliability - quality control charts	(Shewhart)
1932	Transmission electron microscope	(Knoll, Ruska)
1932	Scanning circuit	(Blumlein)
1933	Ignitron	(Westinghouse)
1933	Radio-astronomy	(Jansky)
1932	Hard valve time base circuit	(Puckle)
1933	Frequency-modulation	(Armstrong)
1934	Atomic clock frequency standard	(Cleeton & Williams)
1934	Liquid-crystals	(Dreyer)
1935	Multiplier phototubes	(Zworykin, Morton, Malter)
1935	Travelling wave microwave oscillator	(Heil)
1935	Constant resistance RC circuit	(Blumlein)
1935	Transistor, field effect	(Heil)
1935	Scanning electron microscope	(Knoll)
1936	Vocoder	(Bell Labs.)
1936	Cold cathode trigger tube	(Bell Labs.)
1936	Waveguides	(Southworth, Carson et al)
1936	Long-tailed pair circuit	(Blumlein)
1937	Xerography	(Carlson)
1938	Computers - information theory	(Shannon)
1939	Klystron	(Hahn & Varian Bros.)
1939	Computer - Bell Laboratories	(Stibitz)
1939	Caesium beam frequency standard	(Rabi)
1939	Magnetron	(Randall & Boot)
1939	Digital computer	(Aitken and IBM)
1940	Thick film circuits	(Centralab)
1942	Miller integrator circuit	(Blumlein)
1943	Reliability - sequential analysis	(Wald)
1943	Printed wiring	(Eisler)
1943	Travelling wave tube	(Kompfner, Haeff & Pierce)
1943	ENIAC computer	(Moore school)
1944	Reliability - sampling inspection tables	(Dodge and Romig)
1945	Whirlwind computer	(M.I.T.)
1945	Potted circuits	(Various)
1945	Computer theory	(Von Neuman)
1945	Satellite communication proposal	(Clarke)
1946	CRT storage computer	(Williams)
1947	Chirp radar	(Bell Labs.)
1947	Automatic assembly of radio sets	(Sargrove)
1947	EDVAC computer	(Pennsylvania Univ.)
1947	High quality amplifier circuit	(Williamson)
1947	UNIVAC computer	(Eckert & Maunchly)
1948	Single crystal fabrication - germanium	(Teal & Little)
1948	Communication - information theory	(Shannon)

1948	SEAC computer	(N.B.S.)
1948	Transistor	(Bardeen, Brattain & Shockley)
1948	Film sound recording, magnetic film	(R.C.A. et al)
1948	Holography	(Gabor)
1948	EDSAC computer	(Wilkes)
1949	Cold cathode stepping tube	(Remington Rand)
1949	Dip soldering of printed circuits	(Danko, Abramson)
1949/50	Ion implantation in semiconductors	(Ohl and Shockley)
1950	IBM 650 computer	(I.B.M.)
1950	IBM 701 computer	(I.B.M.)
1950s	Thermo compression bonding	(Anderson, Christensen & Andreatch)
1950s	Ferreed switch	(Bell Labs.)
1951	Quality control	(Juran)
1951	Superregenerative circuit	(Armstrong)
1951	"Tinkertoy" circuit assembly system	(N.B.S.)
1952	Semiconductor integrated circuit concept	(Dummer)
1952	Transistors - zone melting technique	(W.G. Pfann)
1952	Darlington pairs circuit	(Darlington)
1952	SAGE computer	(IBM/MIT)
1952	Single crystal fabrication - silicon	(Teal and Buehler)
1952	Digital voltmeter	(Kay)
1952	Tone control circuit	(Baxendall)
1953	Surface barrier transistor	(Philco)
1953	MASER	(Townes and Weber)
1953	Floating zone silicon refining technique	(Keck, Emeis & Theurer)
1953	Wire wrapping connection technique	(Mallina et al)
1953	IBM 704, 709, 7090 computers	(I.B.M.)
1953	Automatic circuit assembly systems	(General Mills, G.E., Melpar, IBM etc)
1953	Unijunction transistor	(G.E.C.)
1954	Transistor radio set	(Regency)
1954	Solar battery	(Chapin, Fuller, Pearson)
1954	Interdigitated transistor	(Fletcher)
1955	Cryotron	(Buck)
1955	Infra-red emission from Gallium Arsenide semiconductor	(Braunstein)
1956	Diffusion process - silicon	(Fuller & Reis)
1956	Radio Paging	(Multitone)
1956	"Flowsoldering" of printed circuits	(Fry's Metal Foundries)
1956	Vapour cooling of valves	(Beutherel)
1956	YIG material	(Bertaut & Forrat)
1956	Semiconductor diode capacitor	(Giacoletto & O'Connell)
1957	Plated-wire memory	(Gianole)
1957	Scanning electron microscopy	(Oatley et al)
1957	Nickel-chromium film resistors	(Alderton & Ashworth)
1957	Silicon oxide masking process	(Frosch)
1958	Video tape recorder	(Ampex)
1958	Tunnel diode	(Esaki)
1958	LASER	(Schalow & Townes)
1958	Micromodule assembly system	(U.S. Army Signals Estab.)

1958	Mossbauer effect	(Mossbauer)
1958	FET -- Tecnetron	(Tesner)
1958	Field effect varistor	(Bell Labs)
1959	Semiconductor integrated circuit patent	(Kilby)
1959	Tantalum film circuitry	(Bell Labs)
1959	Intrinsic 10μ photo conductor	(Lawson et al)
1959	Planar process	(Hoerni)
1960	Printed wiring multilayer boards	(Photocircuits)
1960	Epitaxial crystal growth (Vapour phase)	(Loor, Christensen et al)
1960	Light emitting diode (LED)	(Allen & Gibbons)
1960	Honeywell 800 computer	(Honeywell)
1960	UNIVAC 80/90 Computer	(I.B.M.)
1960	CD 1604 Computer	(Control Data Corpn.)
1960	Sub-millimetre photoconductive detector	(Putley)
1960	Linear integrated circuits	(Various)
1960	Logic integrated circuits	(Various)
1961	Transferred electron effect	(Ridley & Watkins)
1961	Electronic clock	(Vogel)
1961	Transferred electron device	(Hilsum)
1961	Multilayer seeded plated through hole, printed circuit boards	(Hazeltine)
1961	Epitaxy (Liquid phase)	(Nelson)
1962	Electronic watch	(Vogel)
1962	Semiconductor laser	(Hall et al - also Nathan & Lasher)
1962	Satellite Telstar I	(Various)
1962	Light emitting diode (LED) (Gallium Arsenide phosphide)	(Holonyak)
1962	Flat-pack integrated circuit assembly	(Tao)
1963	Gunn diode oscillator	(Gunn)
1963	Silicon-on-sapphire technology	(Various)
1964	DIP-(DIL)-dual in line package	(Rogers)
1964	IMPATT diode	(Johnston and De Loach)
1964	Overlay transistor	(R.C.A.)
1964	Beam Lead connection technique	(Lepselter)
1965	Self-scanned integrated photodiode arrays	(Weckler)
1965	Minicomputer	(Digital Equip.)
1965	Satellite - Intelstat I	(Various)
1966	Flip-chip bonding technique	(Wiessenstern, Wingrove)
1966	Optical fibres	(Kao and Hockham)
1967	Laser trimming of thick film resistors	(Various)
1967	Surface acoustic wave devices	(White et al)
1968	Amorphous semiconductor switches	(Ovshinsky)
1968	BARITT diode	(Wright)
1969	Magnetic bubbles	(Bobeck, Fischer et al)
1969	Bucket-brigade delay circuit	(Sangster & Teer)
1969	Collector diffusion isolation process	(Bell, Ferranti)
1970	Charge coupled devices	(Boyle & Smith)
1970	X Ray Lithography	(Feder et al)
1971	Oxide defects study by liquid crystals	(Keen)

1972	1024 bit Random Access Memory	(Intel)
1972	Automatic crystal growth control	(Bardsley et al)
1972	X Ray Scanner	(E.M.I.)
1972	V-MOS technique	(Rodgers)
1974	16 bit single chip microprocessor	(National)
1975	Thin film copper bonded process	(Burgess, Neugebauer et al)
1975	4096 6 bit Random Access Memory	(Fairchild)
1975	Integrated optical circuits	(Reinhart, Logan)
1975	Silicon anodisation	(Cook)
1976	Silicon-gate versatile arrays	(Philips)
1976	One board 1/0 computer	(Intel)
1976	16, 384 bit Random Access Memory	(Intel)

Chapter 3. List of inventions by subject

N.B. The date can be used as an approximate index as Chapter 5 is
arranged in chronological order.

1. Assembly techniques and packaging
2. Circuitry
3. Communications
4. Components
5. Computers
6. Industrial

7. Microelectronics
8. Physics
9. Radar
10. Sound reproduction
11. Television
12. Transistors & Semiconductor devices

13. Valves, Tubes, Lamps etc.

* * * * *

1. ASSEMBLY TECHNIQUES AND PACKAGING

1940	Thick film packaged circuits	(Centralab)
1943	Printed wiring	(Eisler)
1945/50	Potted circuits	(Various)
1947	Automatic circuit making equipment	(Sargrove)
1949	Dip soldering of printed circuits	(Danko & Abramson)
1950s	Thermo compression bonding techniques	(Anderson, Christensen & Andreatch)
1951	"TINKERTOY" automatic assembly system	(Nat. Bur. Standards)
1953	Wire wrapping of connections	(Mallina et al)
1953	Automatic assembly systems -- Autofab, Mini-mech, United Shoe Machine, etc.	(General Mills, IBM et al)
1956	"Flowsoldering" of printed circuits	(Fry's Metal Foundries)
1958	"Micro-module" assembly system	(U.S. Army)
1960	Printed wiring multilayer boards	(Photocircuits)
1961	Printed wiring multilayer boards with seeded plated-through holes	(Hazeltine)
1962	"Flat-Pack" integrated circuit	(Tao)
1964	DIP or DIL (Dual in line package)	(Rogers)
1964	Beam lead connections	(Lepselter)
1966	Flip-chip bonding technique	(Wiessenstern & Wingrove)

2. CIRCUITRY

1826	Ohm's law	(Ohm)
1845	Kirchoff's laws	(Kirchoff)
1912	Regenerative circuit	(de Forest, Armstrong & Langmuir)
1912	Heterodyne & superheterodyne circuits	(Fessenden, Armstrong)
1915	Filter networks	(Campbell, Wagner)
1918	Neutrodyne circuit	(Hazeltine)
1918	Shot effect noise	(Schottky)
1918	Multivibrator circuit	(Abraham & Bloch)
1919	Retarded field microwave oscillator	(Barkhausen & Kurtz)
1919	Flip-flop circuit	(Eccles, Jordan)

1921	Crystal control of frequency	(Cady)
1922	Negative resistance oscillator	(Gill, Morrell)
1924	Linear saw-tooth time base circuit	(Anson)
1925	Johnson noise	(Johnson)
1926	Automatic volume control circuit	(Wheeler)
1927	Negative feedback amplifier	(Black)
1932	Energy consuming scanning circuit	(Blumlein)
1933	Hard valve time base circuit	(Puckle)
1935	Constant RC stand-off circuit	(Blumlein)
1936	Long tailed pair circuit	(Blumlein)
1942	Miller integrated circuit	(Blumlein)
1947	High quality amplifier circuit	(Williamson)
1951	Super regeneration	(Armstrong)
1952	Darlington pair circuit	(Darlington)
1952	Digital voltmeter	(Kay)
1952	Negative feedback tone control circuit	(Baxendall)
1960s	Logic circuits - microelectronics	(Various)
1960s	Linear circuits - microelectronics	(Various)
1969	Bucket-brigade delay circuit	(Sangster & Teer)

3. COMMUNICATIONS

1865	Radio wave propagation	(Maxwell)
1876	Telephone	(Bell)
1887	Aerials	(Hertz)
1890	Coherer	(Branley)
1893	Waveguides - theory	(Thomson)
1896	Wireless telegraphy	(Marconi)
1906	Radio broadcasting	(Fessenden)
1912	Ionospheric propagation	(Eccles)
1915	Single sideband transmission	(Carson)
1918	Ground wave propagation	(Watson)
1921	Short wave radio	(Amateurs)
1921	Crystal control of frequency	(Cady)
1925	Short wave commercial	(van Boetzelean)
1926	Yagi aerial	(Yagi)
1928	Diversity reception	(Beverage, Peterson, Moore)
1928	Frequency standards - quartz clock	(Horton, Morrison)
1929	Microwave communication	(Clavier)
1933	Radio astronomy	(Jansky)
1933	Frequency modulation	(Armstrong)
1934	Frequency standards - atomic clock	(Cleeton, Williams)
1936	Waveguides	(Southworth, Carson et al)
1939	Frequency standards - caesium beam	(Rabi)
1945	Satellite communication theory	(Clarke)
1948	Information theory	(Shannon)
1954	Transistor radio set	(Regency)
1956	Radio paging	(Multitone)
1962	Communications satellite TELSTAR I	(Various)

1965   International satellite INTELSTAT I   (Various)
1966   Optical fibre communications   (Kao, Hockham)

4. COMPONENTS

Batteries:
1800   Volta's pile   (Volta)
1803   Accumulator   (Ritter)
1839   Magnetohydrodynamic generation   (Faraday)
1839   Fuel cell   (Grove)
1860   Lead-acid cell   (Plante)
1868   Dry cell   (Leclanche)
1870   Clark standard cell   (Clark)
1884   Zinc mercuric-oxide cell   (Clarke)
1891   Weston standard cell   (Weston)
1900   Nickel-iron cell   (Edison)
1900   Nickel-cadmium cell   (Junger & Berg)
1954   Solar battery   (Chapin, Fuller & Pearson)

Capacitors:
1745   Leyden jar   (von Kleist et al)
1874   Mica capacitors   (Bauer et al)
1876   Rolled paper capacitors   (Fitzgerald)
1900   Ceramic capacitors   (Lombardi)
1904   Glass tubular capacitors   (Moscicki)
1956   Semiconductor diode junction capacitor   (Giacoletto, O'Connell)

Galvanometers:
1820   Electro-magnetism (galvanometer)   (Oersted)
1828   Moving coil   (Schweigger)
1828   Astatic   (Nobilli)

Inductors:
1772   Iron dust cores   (Knight)
1909   Ferrites   (Hilpert, Snoek)

Motors:
1837   Electric motor   (Davenport)
1888   Induction motor   (Tesla)
1902   Synchronous induction motor   (Danielson)

Relays:
1837   Telegraph bell and signal   (Cooke, Wheatstone & Davy)
1950s   Ferreeds   (Bell Labs)

Rheotome:
1868   Waveform plotter   (Lenz)

Resistors:
1850   Thermistor   (Faraday)
1885   Moulded carbon composition   (Bradley)
1897   Carbon film   (Gambrell et al)
1913   Sputtered metal film   (Swann)
1919   Spiralled metal film   (Kruger)

1925	Cracked-carbon	(Siemens and Halske)
1926	Sprayed metal film	(Loewe)
1931	Oxide film	(Littleton)
1957	Nickel-chromium film	(Alderton & Ashworth)
1958	Field effect varistor	(Bell Labs)
1959	Tantalum film	(Bell Labs)

Switches:

1884	Quick break	(Holmes)
1887	Quick make and break	(Holmes)
1950s	Ferreed switch	(Bell Labs)

Transformers:

1831	Transformer	(Faraday)
1885	Distribution	(Deri)
1885	Power	(Zipernowski, Deri & Blathy)

Waveguides:

1893	Theory	(Thomson)
1936	Waveguides	(Southworth, Carson et al)

5. COMPUTERS

1642	Calculating machine	(Pascal)
1672	Calculating machine	(Leibniz)
1833	Calculating machine	(Babbage)
1854	Calculating machine	(Scheutz)
1889	Tabulating machine	(Hollerith)
1931	Differential analyser	(Bush)
1938	Information theory	(Shannon)
1939	Bell Labs complex computer	(Stibitz et al)
1939	Digital computer	(Aitken)
1943	ENIAC	(Moore School)
1945	Whirlwind	(M.I.T.)
1945	Computer theory	(von Neuman)
1946	CRT storage computer	(Williams)
1947	EDVAC	(Penn. Univ.)
1947	UNIVAC	(Eckert, Maunchly)
1948	SEAC	(N.B.S.)
1948	EDSAC	(Wilkes)
1950	IBM 650	(I.B.M.)
1950	IBM 701	(I.B.M.)
1952	SAGE	(I.B.M., M.I.T.)
1953	IBM 704, 709 & 7090	(I.B.M.)
1957	Plated wire memories	(Gianole)
1960	Honeywell 800	(Honeywell)
1960	UNIVAC solid state 80/90	(I.B.M.)
1960	CD 1604	(Control Data Corp.)
1965	Minicomputer	(Digital Equip. Co.)
1969	Magnetic bubble memories	(Bobeck et al)
1970	Charge coupled device memories	(Boyle, Smith)

1972	1024 bit random access memory	(Intel)
1974	16 bit single chip microprocessor	(National)
1975	16,384 bit random access memory	(Intel)
1976	One board computer with programmable 1/0	(Intel)

6. INDUSTRIAL

1839	Microfilming	(Dancer)
1843	Facsimile reproduction	(Bain)
1908	Geiger counter	(Geiger, Rutherford)
1912	Tunger rectifier	(Langmuir)
1913	Reliability standards	(AIEE)
1914	Ultrasonics	(Longevin)
1914	Thyratrons	(Langmuir)
1916	Reliability, control system	(Bell/Western Electric)
1918	Induction heating	(Northrup)
1920	Ultra-micrometer	(Whiddington)
1931	Reliability - quality control-charts	(Shewhart)
1933	Ignitron	(Westinghouse)
1943	Reliability - sequential analysis	(Wald)
1944	Reliability - sampling inspection tables	(Dodge & Romig)
1951	Quality control	(Juran)
1961	Electronic clock	(Vogel et Cie)
1962	Electronic watch	(Vogel et Cie)

7. MICROELECTRONICS
(Refer also to 12 - Transistors & semiconductor devices)

1852	Thin films sputtering process	(Grove)
1913	Sputtered metal film resistors	(Swann)
1940	Thick film circuits	(Centralab)
1949/50	Ion implantation	(Ohl & Shockley)
1952	Semiconductor integrated circuit concept	(Dummer)
1952	Zone melting technique	(Pfann)
1957	Nickel chromium thin film resistors	(Alderton, Ashworth)
1959	Semiconductor integrated circuit patent	(Kilby)
1959	Tantalum thin film circuits	(Bell Labs)
1959	Planar process	(Hoeni)
1960	Epitaxy - vapour phase	(Loor, Christensen et al)
1960	Digital & linear integrated circuits	(Various)
1961	Epitaxy - liquid phase	(Nelson)
1965	Minicomputer	(Digital Equip. Co.)
1967	Laser trimming of thick film resistors	(Various)
1967	Surface acoustic wave devices	(White et al)
1969	Collector diffusion isolation	(Bell Labs., Ferranti)
1970	X Ray lithography	(Feder et al)
1971	Liquid crystal study of oxide defects	(Keen)
1972	1024 bit random access memory	(Intel)
1972	V-MOS technique	(Rodgers)
1974	16 bit single chip microprocessor	(National)
1975	Thin film direct bonded copper process	(Burgess, Neugebauer et al)

1975	Integrated optical circuits	(Reinhart, Logan)
1975	4096 bit random access memory	(Fairchild)
1975	Silicon anodisation	(Cook)
1976	Microelectronic versatile arrays	(Philips)
1976	One board with programmable 1/0 computer	(Intel)
1976	16,384 bit random access memory	(Intel)

8. PHYSICS

1780	Galvanic action	(Galvani)
1800	Infra-red radiation	(Herschel)
1801	Ultra-violet radiation	(Ritter)
1808	Atomic theory	(Dalton)
1820	Electro-magmetism	(Oersted)
1821	Thermoelectricity	(Seebeck)
1826	Ohm's law	(Ohm)
1831	Electromagnetic induction	(Faraday)
1832	Self induction	(Henry)
1834	Electrolysis	(Faraday)
1839	Photovoltaic effect	(Becquerel)
1847	Magnetostriction	(Joule)
1851	Relation between theory of magnetism and electricity	(Kelvin)
1860	Electron	(Thomson)
1878	Cathode rays	(Crookes)
1879	Hall effect	(Hall)
1880	Piezo electricity	(Curies)
1882	Wimshurst machine	(Wimshurst)
1895	X-rays	(Rontgen)
1897	Cathode ray oscillograph	(Braun)
1912	Cloud chamber	(Wilson)
1921	Ferroelectricity	(Vasalek)
1929	Cyclotron	(Laurence)
1930	Van de Graaf accelerator	(Van de Graaf)
1932	Transmission electron microscope	(Knoll, Ruska)
1934	Liquid crystals	(Dreyer)
1935	Scanning electron microscope	(Knoll)
1937	Xerography	(Coulson)
1948	Holography	(Gabor)
1953	MASER	(Townes & Weber)
1955	Infra-red emission from GaSb	(Braunstein)
1955	Cryotron	(Buck)
1956	YIG magnetic materials	(Bertaut, Forrat)
1957	Scanning electron microscopy	(Oatley et al)
1958	LASER	(Schalow, Townes)
1958	Mossbauer effect	(Mossbauer)
1959	Intrinsic 10μ photoconductor	(Lawson et al)
1960	Sub-millimetre photoconductive detector	(Putley)
1961	Transferred electron effect	(Ridley, Watkins)
1961	Transferred electron device	(Hilsum)

1962	Semiconductor laser	(Hall et al - also Nathan and Lasher)
1962	LED (Gallium arsenide phosphide)	(Holonyak)
1963	Gunn diode oscillator	(Gunn)
1967	Surface acoustic wave devices	(White et al)
1970	X-ray lithography	(Feder et al)
1972	X-ray scanner	(E.M.I.)
1972	Automatic crystal growth control	(Bardsley et al)

9. RADAR

1924	Radar systems	(Appleton, Briet, Watson Watt et al)
1947	Chirp technique	(Bell Labs)

10. SOUND REPRODUCTION

1860	Microphone, diaphragm type	(Reis)
1876	Telephone	(Bell)
1877	Phonograph	(Edison)
1877	Microphone, carbon	(Edison)
1877	Loudspeaker, moving coil	(Siemens)
1887	Gramophone	(Berliner)
1898	Magnetic recording	(Poulsen)
1908	Electronic organ	(Cahill)
1917	Microphone, condenser	(Wente)
1925	Loudspeaker, electrostatic	(Various)
1926	Films, sound-on-disc system	(Warner Bros)
1927	Films, sound-on-film system	(Fox Movietone News)
1930s	Radiophonic sound and music	(Grainger)
1931	Stereophonic sound reproduction	(Blumlein, Bell Labs)
1936	Vocoder	(Bell Labs)
1948	Films - magnetic recording	(RCA et al)
1958	Video tape recorder	(Ampex)

11. TELEVISION

1908	Electronic system - theory	(Campbell-Swinton)
1919	Electronic system	(Zworykin)
1923	Iconoscope	(Zworykin)
1925	Mechanical system	(Baird)
1927	Electronic system (see 1919)	(Farnsworth)

12. TRANSISTORS & SEMICONDUCTOR DEVICES
(Refer also to 7 - Microelectronics)

1917	Crystal pulling process	(Czochralski)
1930	MOS/FET	(Lilienfeld)
1935	Field effect transistor	(Heil)
1948	Single crystal fabrication - germanium	(Teal and Little)
1948	Transistor	(Bardeen, Brattain, Shockley)
1949/50	Ion implantation	(Ohl and Shockley)
1952	Zone melting technique	(Pfann)
1952	Single crystal fabrication - silicon	(Teal, Bueler)
1953	Surface barrier transistor	(Philco)

1953	Floating zone refining process	(Keck, Emeis & Theurer)
1953	Unijunction transistor	(G.E.C.)
1954	Transistor radio set	(Regency)
1954	Silicon solar battery	(Chapin, Fuller & Pearson)
1954	Interdigitated transistor	(Fletcher)
1955	Infra-red emission from GaSb	(Braunstein)
1956	Diffusion process	(Fuller, Reis)
1956	Semiconductor diode junction capacitor	(Giacoletto & O'Connell)
1957	Oxide masking process	(Frosch)
1958	Tunnel diode	(Esaki)
1958	"Technetron" FET	(Teszner)
1958	Field effect varistor	(Bell Labs)
1959	Planar process	(Hoerni)
1960	Light emitting diode (LED)	(Allen & Gibbons)
1960	Epitaxy (Vapour phase)	(Loor, Christensen et al)
1961	Epitaxy (Liquid phase)	(Nelson)
1962	LED (Gallium ArsenidePhosphide)	(Holonyak)
1963	Gunn diode oscillator	(Gunn)
1963	Silicon-on-sapphire technology	(Various)
1964	IMPATT diode	(Johnston, de Loach)
1964	Overlay transistor	(R.C.A.)
1965	Self-scanned photodiode arrays	(Weckler)
1968	Amorphous semiconductor switches	(Ovshinsky)
1968	BARRITT diode	(Wright)
1969	Magnetic bubbles	(Bobeck, Fischer et al)
1970	Charge coupled devices	(Boyle, Smith)
1970	X ray lithography for bubble devices	(Feder et al)

13. VALVES, TUBES, LAMPS etc.

1855	Cold cathode discharge tube	(Gaugain)
1856	Low pressure discharge tube	(Geissler)
1857	Mercury arc lamp	(Way)
1878	Carbon filament lamp	(Swan, Stearn et al)
1901	Fluorescent lamp	(Cooper-Hewitt)
1904	Two electrode valve	(Fleming)
1906	Three electrode valve	(de Forest)
1910	Neon lamp	(Claude)
1912	Tungar rectifier	(Langmuir)
1914	Thyratron	(Langmuir)
1919	Retarded field microwave oscillator	(Barkhausen, Kurtz)
1919	Houskeeper seal	(Houskeeper)
1922	Negative resistance oscillator	(Gill, Morrell)
1926	Screened grid valve	(Round)
1928	Pentode valve	(Tellegen, Holst)
1933	Ignitron	(Westinghouse)
1935	Travelling wave microwave oscillator	(Heil)
1935	Multiplier phototube	(Zworykin, Morton & Malter)
1936	Cold cathode trigger tube	(Bell Labs)

1939	Klystron	(Hahn and Varian Bros)
1939	Magnetron	(Randall and Boot)
1943	Travelling wave tube	(Kompfner et al)
1949	Cold cathode stepping tube	(Remington Rand)
1956	Vapour cooling of valves	(Beutheret)

Chapter 4.　List of inventors

(In alphabetical order, with dates of inventions)

Abraham H. 1918

Aitken H.H. 1939

Alderton R.H. 1957

Allen J.W. 1960

Anderson O.L. 1950s

Andreatch P. 1950s

Anson R. 1924

Appleton E. 1924

Armstrong E.H. 1912, 1912, 1933, 1951.

Ashworth F. 1957

Babbage C. 1833

Bacon F.T. 1839

Bain A. 1843

Baird J.L. 1925

Bardsley W. 1972

Bauer M. 1874

Baxendall P.J. 1952

Bardeen J. 1948

Barkhausen H. 1919

Becquerel E. 1839

Bell A.G. 1876

Berg 1900

Berliner E. 1887

Bertout F. 1956

Beverage H.A. 1928

Block H.S. 1927

Bloch E. 1918

Blumlein A.D. 1931, 1932, 1935, 1936, 1942.

Bobeck A.H. 1969

Boetzelaer, van 1925

Boot H.A.H. 1939

Boyle W.S. 1970

Bradley C.S. 1885

Branley E. 1890

Brattain W.H. 1948

Braun F. 1897

Braunstein R. 1955

Briet O. 1924

Buck D. 1955

Buehler E. 1952

Burgess J.F. 1975

Bush V. 1931

Butheret C. 1956

Cady W.G. 1921

Cahill T. 1908

Campbell G. 1915

Campbell-Swinton A.A. 1908

Carlson C. 1937

Carlson R.O. 1962

Carson J.R. 1915, 1936

Chapin D.M. 1954

Christensen H. 1950s, 1960

Clark L. 1870

Clarke A.C. 1945

Clarke C.L. 1884

Claude G. 1910

Clavier A.G. 1929

Cleeton C.E. 1934

Cook R. 1975

Cooke W.F. 1837

Cooper-Hewitt P. 1901

Crookes W. 1878

Cross C.F. 1878

Cunaeus 1745

Curie J. 1880

Curie P. 1880

Czochralski J. 1917

Dalton J. 1808

Dancer 1839

Danielson E. 1902

Danko S.F. 1949

Darlington S. 1952

Davenport T.

Davy E. 1837

De Loach B.C. 1964

Dodge H.F. 1944

Dreyer J. 1934

Dummer G.W.A. 1952

Eccles 1919

Eccles W.H. 1912

Eckert P. 1947

Edison T.A. 1877, 1877, 1900.

Eisler P. 1943

Emeis R. 1953

Esaki L. 1958

Faraday M. 1831, 1831, 1834, 1839, 1850.

25

Farnsworth P. 1919

Feder E. 1970

Fenner G.E. 1962

Fessenden R. 1906, 1912

Fischer R.F. 1969

Fitzgerald D.G. 1876

Flanagan G. 1975

Fleming J.A. 1904

Fletcher N.H. 1954

Forest de 1906, 1912

Forret F. 1956

Frosch E.J. 1957

Fuller C.S. 1954, 1956

Gabor D. 1948

Galvani L. 1780

Gambrell T.E. 1897

Gaugain J.M. 1855

Geiger H. 1908

Giacoletto L.J. 1956

Gianole V.F. 1957

Gibbons P.E. 1960

Geissler H. 1856

Gill E.W.B. 1922

Graaf van de .1930

Grainger P. 1930s

Green G. W. 1972

Grove W.R. 1839, 1852

Gunn J.B. 1963

Haeff A.W. 1943

Hall E.H. 1879

Hall R.N. 1962

Hahn W.C. 1939

Harris A.F. 1897

Hazeltine L.A. 1918

Heil A. 1935

Heil O. 1935, 1935

Henry J. 1832

Herschel 1800

Hertz H.R. 1887

Hilpert G. 1909

Hockham G.A. 1966

Hoeni J.A. 1959

Hollerith H.1889

Holliday E.H. 1972

Holmes J.H. 1884, 1887

Holonyak N. 1962

Holst 1928

Horton J.W. 1928

Houskeeper W.G. 1919

Hurle D.T.J. 1972

Jansky K.G. 1933

Johnson J.B. 1925

Johnston R.L. 1964

Jordan 1919

Joule J. 1847

Junger 1900

Juron J.M. 1951

Kao K.C. 1966

Kay A. 1952

Keck P.H. 1953

Keen J.M. 1971

Kelvin, Lord 1851

Kilby J.S. 1959

Kingsley J.D. 1962

Kirchoff G.R. 1845

Kleist von 1745

Kleimock J.J. 1959

Knight G.1972

Knoll M. 1932, 1935

Kompfner R. 1943

Kruger F. 1919

Kurtz K. 1919

Langevin P. 1914

Langmuir I. 1912, 1912, 1914

Lasher G. 1962

Laurence E.O. 1929

Lawson W.D. 1959

Leclanche G. 1868

Leibniz G.W. 1672

Lenz H. 1868

Lepselter M. 1964

Logan R.A. 1975

Lilienfeld J. 1930

Little J.B. 1948

Littleton J.T. 1931

Loewe S. 1926

Lombardi L. 1900

Loor H.H. 1959

Luc, de 1800

Mallina R.F. 1953

Malter L. 1935

Marconi G. 1896

Morrison W.A. 1928

Maunchly J. 1947

Maxwell J.C. 1865

Meissner 1912

Moore J.B. 1928

Moore R.E. 1975

Morrell J.H. 1922

Morton G.A. 1935

Moscicki I. 1904

Mossbauer R.L. 1958

Mushenbrook, von 1745

Nathan M.I. 1962

Nelson H. 1961

Neugebauer C.A. 1975

Neumann, von 1945

Nielsen S. 1959

Nobilli 1828

Northrup E.F. 1918

Oatley C.W. 1957

O'Connell J. 1956

Oersted H.C. 1820

Ohm G.S. 1826

Ohl R.S. 1949/50

Ovshinsky S.R. 1968

Pascal B. 1642

Poulsen V. 1898

Perneski A.J. 1969

Pearson G.L. 1954

Peterson H.O. 1928

Pfann W.G. 1952

Pierce J.R. 1943

Plante 1860

Puckle O.S. 1933

Putley E.H. 1959, 1960

Rabi I.I. 1939

Randall J.T. 1939

Reinhart F.K. 1975

Reis H. 1956

Reis J.P. 1860

Remeika J.P. 1969

Ridley B.K. 1961

Ritter J.W. 1801, 1803

Rodgers T.J. 1972

Rogers B. 1964

Romig H.C. 1944

Rontgen W.K. 1895

Round H.J. 1926

Ruska E. 1932

Rutherford E. 1908

Sangster F.J.L. 1969

Sargrove J.A. 1947

Schottky W.

Schweigger J. 1828

Seebeck T.J. 1821

Shannon C.E. 1938, 1948

Schelkunoff S.A. 1936

Schalow A.L. 1958

Scheutz P.G. 1854

Shewhart W.A. 1931

Shockley W. 1948, 1949/50

Siemens E.W. 1877, 1925

Smith G.E. 1970

Snoek J.L. 1909

Soltys T.J. 1962

Southworth G.C. 1936

Stearn C.H. 1878

Stibitz G. 1939

Swan J.W. 1878

Swann W.F.G. 1913

Tao Y. 1962

Teal G.K. 1948, 1952

Teer K. 1969

Tellegen 1928

Tesla N. 1888

Teszner S. 1958

Theurer H.C. 1953, 1960

Thomson J.J. 1860, 1893

Topham F. 1878

Townes C.H. 1953, 1958

Varian S.F. 1939

Varian R.H. 1939

Vasalek J. 1921

Vitert L.G. van 1969

Vogel P. 1961, 1962

Volta A. 1800

Wagner K.W. 1915

Wald A. 1943

Warner Bros. 1926

Way J.T. 1857

Watkins T.B. 1961

Watson G.N. 1918

Watson-Watt R.A. 1924

Weber J. 1953

Weckler G.P. 1965

Wente E.C. 1917

Weston E. 1891

Wheatstone 1837

Wheeler H.A. 1926

Whiddington R. 1920

White R.M. 1967

Wiessenstern M. 1966

Williams F. C. 1946

Williams N. A. 1934

Williamson D. T. N. 1947

Wilkes M. V. 1948

Wilson C. T. R. 1912

Wimshurst J. 1882

Wingrove G. A. S. 1966

Wright G. T. 1968

Yagi H. 1926

Young A. S. 1959

Zamboni 1800

Zworykin V. K. 1919, 1923, 1935.

Chapter 5.   Brief description of each invention, in date order

1642   <u>COMPUTER (Mechanical Calculating</u>   Blaise Pascal (France)
       <u>Machines)</u>.

<u>EDITORS NOTE</u>:- Although non-electronic this item is included as an
essential part of computer history.

       The invention of the first mechanical device capable of addition and
subtraction in a digital manner has been generally credited to Pascal, who
built his first machine in 1642. This claim has been contested on the basis of
letters sent to Kepler in 1623 and 1624 by Wilhelm Schickhardt of Tubingen, in
which the latter describes the construction of a calculator. Pascal, who at the
age of 19 had wearied of adding long columns of figures in his father's tax office
in Rouen, made a number of calculators, some of which are still preserved in
museums. His machines had number wheels with parallel, horizontal axes.
The positions of these wheels could be observed and sums read through windows
in their covers. Numbers were entered by means of horizontal telephone-dial-
like wheels which were coupled to the number wheels by pin gearing. Most of
the number wheels were geared for decimal reckoning but the two wheels on
the extreme right had twenty and twelve divisions, respectively for sous and
deniers. A carry ratchet coupled each wheel to the next higher place. The
stylus-operated pocket adding machines now widely used are descendents of
Pascal's machine.

<u>SOURCE</u>: Serrell, Astrahan, Patterson and Pyne "The evolution of computing
machines and systems" Proc. IRE May 1962. p.1041.

<u>SEE ALSO</u>: "The Inventor of the First Desk Calculator" V.P. Czapla.
"Computers and Automation" Vol.10. p.6. September 1961.

<u>ALSO</u>: "The computer from Pascal to von Neuman" by H.H. Goldstine.
Princeton Univ. Press 1972. p.7.

<u>ALSO</u>:   "The origins of digital computers" Edited by B. Randell. Springer-
Verlag, Berlin 1973. p.2.

1672   <u>COMPUTERS (Mechanical</u>   G.W. Leibniz (Germany)
       <u>Calculating Machines)</u>

<u>EDITORS NOTE</u>:- Although non-electronic, this item is included as part of
computer history.

       Gottfried Wilhelm Leibniz invented the "Leibniz Wheel" which enabled
him to build a calculating machine which surpassed Pascal's in that it could do,
not only addition and subtraction fully automatically but also multiplication and
division.

<u>SOURCE</u>: "The computer from Pascal to von Neumann" by H.H. Goldstine.
Princeton Univ. Press 1972. p.7.

<u>SEE ALSO</u>: "The origins of digital computers" Edited by B. Randell. Springer-
Verlag, Berlin.1973. p.2.-   29

1745   <u>CAPACITOR (LEYDEN JAR)</u>   von Mushenbrock and Cunaeus (Germany)
                                                        and
                                             von Kleist (Pomerania)

       According to the literature, the Leyden Jar was discovered almost simultaneously by Dean von Kleist of the Cathedral of Camin, Germany, in October 1745, and Peter von Muschenbrock, Professor in the University of Leyden, in January 1746, over 200 years ago. As described by them, it was a glass jar or vial with inner and outer electrodes of various things - water, mercury, metal foil, etc. The modern miniature glass dielectric capacitor differs in form and structure from the 200-year-old Leyden Jar, but the principle of operation is the same.

SOURCE: "History, Present Status and Future Developments of Electronic Components" by P.S. Darnell. IRE Transactions on Component Parts. September 1958. p. 127/8

SEE ALSO: "Janus" C. Dorsman and C.A. Crommelin. 46, 1957. p. 274.

ALSO: "Observations on the manner in which glass is charged with electric fluid" by E.W. Gray. Phil. Trans. Royal Soc. London Vol. 77 (1788) p. 407.

ALSO: "Residual charge of the Leyden jar - dielectric properties of various glasses" by J. Hopkinson. Phil. Mag. Part 5. Vol. 4 (1877) p. 141.

1772   <u>IRON DUST CORES</u>   Gavin Knight (U.K.)

       Iron dust cores consisting of iron filings churned in water, bound with linseed oil, moulded and fired. They were used in a Navy compass. Apparently, Knight was a secretive person and details of his process were not actually published before 1779.

SOURCE: Note from British Science Museum.

SEE ALSO: Andrade -'The early history of the permanent magnet.'Endeavour. January 1958. p. 27.

ALSO: Benjamin Wilson "Phil Trans" Vol. 69, 1779. p. 51. (giving details of the process).

1780   <u>GALVANIC ACTION</u>   L. Galvani   (Italy)

       Luigi Galvani began his studies on the subject of animal electricity in 1780. When performing experiments on nervous excitability in frogs, he saw that violent muscle contractions could be observed if the lumbar nerves of the frog were touched with metal instruments in the presence of distant electrical discharges.

       The word "electricity" was reserved for static electricity and the word "Galvanism" was proposed by von Humboldt for direct (continuous) current.

SOURCE: "From torpedo to telemetry" by D.W. Hill. Electronics & Power. 27th November, 1975. pp. 1110-1111.

1800   **DRY BATTERY**   A. Volta (1800) De Luc (1809)
                                        and Zamboni (1812)   (Italy)

     Volta's invention of the electric battery was announced in a letter to Sir Joseph Banks, the President of the Royal Society and described his "Volta's Pile" - consisting of copper and zinc discs separated by a moistened cloth electrolyte. These were later improved to consist of paper discs tinned one side, manganese dioxide on the other, stacked to produce 0.75 v. between 1 in. diam. discs.

Note by Science Museum, London:-
     Scyffer described experiments with dry cells carried out by Ludicke (1801) Einhof, Ritter (1802) Hachette and Desornes, Biot, and many others. Scyffer regarded these as experimental and ascribes the first effective pile to Behrens and to Marechaux but considered the best performance before Zamboni to have been achieved by De Luc in 1809. Zamboni (1812) himself ascribed priority to De Luc since his paper was entitled: "Descrizione della colonna elettrica del Signore de Luc e considerazione sull analisi de lui Fatta della pile Voltiana". Work on early dry batteries was, therefore, done from about 1800 to 1812.

SOURCE: "On the electricity excited by the mere contact of conducting substances of different kinds" Phil. Trans. Vol. 90. (1800) page 403.

SEE ALSO: "A Biographical Dictionary of Scientists" T.I. Williams. p. 535 (Volta) Adam and Charles Black, London 1969.

ALSO: Scyffer "Geschichtliche Dartellung der Galvanisms" 1848. p. 135-148.

1800   **INFRA-RED RADIATION**   W. Herschel   (U.K.)

     In 1800, William Herschel, during research into the heating effects of the visible spectrum, discovered that the maximum heating was not within the visible spectrum but just beyond the red range. Herschel concluded that in addition to visible rays the sun emits certain invisible ones. These he called infra-red rays.

SOURCE: "Electronics Engineer's Reference Book" Newnes-Butterworth London (1976) Chap. 4. p. 4 - 2.

1801   **ULTRA-VIOLET RADIATION**   J. W. Ritter (Germany)

     In 1801, the German physicist Ritter made a further discovery. He took a sheet of paper freshly coated with Silver Chloride and placed it on top of a visible spectrum produced from sunlight falling through a prism. After a while he examined the paper in bright light. It was blackened, and it was most blackened just beyond the violet range of the spectrum. These invisible rays Ritter called ultra-violet rays.

SOURCE: "Electronics Engineer's Reference Book" Newnes-Butterworth London (1976) Chap. 4. p. 4 - 2.

1803   ACCUMULATOR   J. W. Ritter (Germany)

Ritter's charging or secondary pile consists of but one metal, the discs
of which are separated by circular pieces of cloth, flannel or cardboard, moist-
ened in a liquid which cannot chemically affect the metal. When the extremities
are put in communication with the poles of an ordinary voltaic pile it becomes
electrified and can be substituted for the latter and it will retain the charge.

SOURCE: "Biographical History of Electricity and Magnetism" Mottelay.
Charles Griffin & Co. London 1922. p. 381.

1808   ATOMIC THEORY   J. Dalton (U.K.)

Dalton conceived the idea that the atoms of different elements were
distinguished by differences in their weights.   In 1808, he propounded the
theory that all chemical elements are composed of minute particles of matter
called atoms and that these atoms, as the name implies, cannot be cut up any
further. All atoms of one element, he said, were alike but atoms of different
elements had different weights. The atom of hydrogen was the lightest atom
$(1.66 \times 10^{-24}$ of a gram) and the weights of all other atoms were compared with
it.

SOURCE: "New System of Chemical Philosophy" J. Dalton 1808.

SEE ALSO: "A Biographical Dictionary of Scientists" T. I. Williams. Adam
and Charles Black, London 1969. p. 128.

1820   ELECTRO-MAGNETISM   H. C. Oersted   (Denmark)
        (Galvanometer)

In 1820, Oersted reported the discovery of electro-magnetism, and this
led him to develop the first galvanometers. It was John Schweigger who constructed the first moving-coil instrument and Nobilli (1828) an Italian physicist,
developed a sensitive astatic galvanometer and compared its sensitivity with
that of the most 'sensitive galvanometer' then available.

SOURCE: "From torpedo to telemetry" by D. W. Hill. Electronics & Power.
27th November, 1975. p. 111.

SEE ALSO: "Experiments on the effect of a current of electricity on the
magnetic needle" by H. C. Oersted. Annals of Philosophy. Vol. 16. London
(1820) p. 273.

1821   THERMOELECTRICITY   T. J. Seebeck (Germany)

The discovery of thermoelectricity is usually attributed to Professor
T. J. Seebeck of Berlin in 1821 although there is some evidence that he might
have been anticipated by Dessaignes in 1815. Professor Cummings of Cambridge also discovered the effect independently and published his findings in
1823.

Following Seebeck's work, J. C. A. Peltier completed Seebeck's discovery
by showing that the passage of electricity through a junction of two different
metals (antimony and copper) could produce a rise in temperature at the

junction when passing in one direction and a drop in temperature when passing
in the contrary direction.

    The introduction of the first successful thermopile in the sense of an
array of thermocouples (analogous to the galvanic pile) is attributed to Nobilli.
Nobilli's thermopile was subsequently improved by Melloni.

SOURCE: Note from British Science Museum, London (1st and 3rd paragraphs)
Author (2nd paragraph).

1826   OHMS LAW   G.S.Ohm   (Germany)

    Ohm was Head of the Department of Mathematics and Physics at the
Polytechnic Institute in Cologne when he discovered the law bearing his name.

$$E = IR \qquad R = \frac{E}{I} \qquad I = \frac{E}{R}$$

1828   MOVING COIL GALVANOMETER   J.Schweigger   (Germany)
1828   ASTATIC GALVANOMETER   C.L.Nobilli   (Italy)

    In 1820, Oersted reported the discovery of electro-magnetism, and this
led him to develop the first galvanometers. It was John Schweigger who constructed the first moving-coil instruments and Nobilli (1828) an Italian physicist,
developed a sensitive astatic galvanometer and compared its sensitivity with
that of the most 'sensitive galvanometer' then available.

SOURCE: "From torpedo to telemetry" by D.W.Hill.   Electronics & Power.
27th November,1975. p.111.

SEE ALSO: "Comparison entre les deux galvanometres les plus sensibles, la
grenouille et le multiplicateur a deux aiguilles, suivie de querques resultats
nouveaux" C.L.Nobilli. Chim.et Phys. RS (1828) 43, pp.256-258.

1831   ELECTROMAGNETIC   M.Faraday (U.K.)
INDUCTION (see also SELF
INDUCTION 1832

    In 1831 Faraday wound an iron ring with two coils: one connected to a
voltaic battery, was to create the primary vibration - the iron ring was to concentrate the lateral vibrations from this - and another coil on the opposite side
of the ring was to convert these secondary vibrations into another electric
current. Thus, on 29th August 1831, Farday discovered electromagnetic
induction.

SOURCE: "A Biographical Dictionary of Scientists" T.I.Williams. Adam &
Charles Black, London 1969. p.174.

SEE ALSO: "Faraday's Discovery of Electro-Magnetic Induction" by T.Martin
Ed.Arnold & Co.London 1949. p.52.

1831   TRANSFORMER   Michael Faraday (U.K.)
Note by British Science Museum:-

    "The word "transformer" was first used in the electrical sense
in 1883 for both static transformers and rotating machines (motor generators).
Previous to this "induction coil" had been used. Faraday was the first to link

two electric circuits by a magnetic circuit. The earliest application of the
device to transfer power appears to be Jablochkoff's patent No:1996 (date
of application 22nd May 1877). *This was really the first practical "transformer" and it utilized a piece of apparatus which had existed since 1831 ."

A contrivance was used by Michael Faraday in his experiments
on electromagnetic induction. This device is described in his diary under
the date of August 29, 1831:

> "Expts. on the production of Electricity from Magnetism, etc. etc.
> Have had an iron ring made (soft iron), iron round and 7/8 inch
> thick and ring 6 inches in external diameter. Wound many coils
> of copper wire round one half, the coils being separated by twine
> and calico - there were three lengths of wire each about 24 feet
> long and they could be connected as one length or used as separate
> lengths. By trial with a trough each was insulated from the other.
> Will call this side of the ring A. On the other side but separated
> by an interval was wound wire in two pieces together amounting to
> about 60 feet in length, the direction being as with the former
> coils; this side call B.
> Charges a battery of 10 pr. plates 4 inches square. Made the coil
> on B side one coil and connected its extremities by a copper wire
> passing to a distance and just over a magnetic needle (3 feet from
> iron ring). Then connected the ends of one of the pieces on A side
> with battery; immediately a sensible effort on needle. It oscillated
> and settled at last in original position. On breaking connection of
> A side with battery again a disturbance of the needle.
> Made all the wires on A side one coil and sent current from
> battery through the whole. Effect on needle much stronger than
> before."

A multiwinding transformer and a transformer experiment are described by these words of Faradays:-

SOURCE: "History, present status and future developments of electronic
components" by P.S. Darnell. IRE Transactions on Component parts.
September, 1958. p. 125.

SEE ALSO: "Faraday's discovery of electro-magnetic induction" by
T. Martin. Edward Arnold & Co., London, Eng. pp. 52-54. 1949.

*Jablochkoff's patent covered a lighting system which includes a mention of a
transformer transferring power supply to a number of lamps.

1832   SELF INDUCTION (see ELECTROMAGNETIC   J. Henry (U.S.A.)
        INDUCTION 1831)

Henry is reported to have discovered the phenomenom of self induction
in 1830 but, through his failure to publish, priority was given to Michael
Faraday to Henry's great mortification.

REFERENCE: "Joseph Henry" by T. Coulson. Princeton Univ. Press U.S.A. 1950.

SEE ALSO: "A Biographical Dictionary of Scientists" T.I. Williams. Adam and Charles Black, London 1969. p. 250.

1833   COMPUTER (Calculating Machines)   C. Babbage (U.K.)

      In 1833 Babbage conceived his analytical engine, the first design for a universal automatic calculator. He worked on it with his own money until his death in 1871. Babbage's design had all the elements of a modern general - purpose digital computer, namely: memory, control, arithmetic unit and input/output. The memory was to hold 1000 words of 50 digits each, all in counting wheels. Control was to be by means of sequences of Jacquard punched cards. The very important ability to modify the course of a calculation according to the intermediate results obtained - now called conditional branching - was to be incorporated in the form of a procedure for slipping forward or backward a specified number of cards.

SOURCE: Serrell, Astrahan, Patterson and Pyne "The evolution of computing machines and systems" Proc. IRE May 1962. p. 1042.

SEE ALSO: "The computer from Pascal to von Neumann" by H.H. Goldstine. Princeton Univ. Press 1972. p. 10.

ALSO: "The origins of digital computers" Edited by B. Randell. Springer-Verlag, Berlin 1973. p. 7.

ALSO: "On the mathematical powers of the calculating engine" by Charles Babbage, 26th December 1837 (in Randell's book p. 17.)

1834   ELECTROLYSIS   M. Faraday (U.K.)

      Faraday announced his two laws of electrolysis in 1834 which made explicit the amount of force required for a given amount of electrical force, chemical substances in the ratio of their chemical equivalent were released at the electrodes of an electrochemical cell.

1837   RELAYS   W.F. Cooke, Wheatstone and E. Davy (U.K.)

Telegraph Bell Relay
Telegraph Signal Relay

      The first patent was taken out by Edward Davy in 1838 (British Patent No: 7719). "I claim the mode of making telegraph signals or communications from one distant place to another by employment of relays of metallic circuits brought into operation by electric currents". In 1837 (British Patent No:7390) Cooke and Wheatstone described an electromagnetic relay device for bringing a local battery at the distant station into action to sound an alarm bell there. However, Davy was described by Fakie as working on telegraphy as early as 1836 and entered an opposition to Cooke and Wheatstone's 1837 application for a patent, but the patent was granted. Morse in the U.S.A. is credited with a patent in 1840 (U.S. Patent No: 1647) which is apparently similar to Davy's patent.

SOURCE: Note by British Science Museum, London.

SEE ALSO: J.J. Fakie "A History of Electric Telegraphy in the year 1837"

London 1884.

ALSO: "A Biographical Dictionary of Scientists" T.I. Williams. Adam and
Charles Black, London 1969. (Sir W. Fothergill Cooke. p.114).

1837   ELECTRIC MOTOR   T. Davenport (U.S.A.)

The earliest known example of a patent for an electric motor is U.S.
Patent No:132 granted on 25th February, 1837 to Thomas Davenport, of Brandon,
Vermont, entitled: "Improvement in Propelling Machinery by Magnetism and
Electro-magnetism".

According to the description contained in the specification, the motor,
which is intended to be driven by a 'galvanic battery', is constructed on sound
electro-magnetic principles.

SOURCE: "Patents for Engineers" by L.H.A. Carr and J.C. Wood. Chapman
and Hall, London 1959. p. 87.

1839   BATTERY (Magnetohydrodynamic)   M. Faraday (U.K.)

The idea of producing electricity from a moving fluid, which is the
basis of an MHD generator, was proposed by Faraday in 1839. A conducting
fluid is passed between the poles of a magnet and an electromotive force is
produced at right angles to the field. This principle is also used in electro-
magnetic pumps and induction flow meters for conducting fluids. In recent
proposals by Kantrowitz and Spron (1959) the working fluid is a conducting gas
at high temperatures. The gas moving at a high velocity is passed through a
magnetic field at right angles to the direction of flow; electrodes placed on
opposite sides of the channel extract the power and are connected to the external load.

SOURCE: "The magnetohydrodynamic generation of power"   by K. Phillips.
AEI Engineering (Mar/Apr.1964) p. 62.

SEE ALSO: "Experimental researches in electricity" by M. Faraday, London.
1839.

ALSO: "Application of the MHD concept to large scale generation of electric
power" by A. Kontrowitz and P. Sporn. American Electric Power Service
Corporation and AVCO Research Laboratory (1959).

1839   BATTERY (Fuel cell)   W.R. Grove (U.K.)

The fuel cell principle - that is, the conversion of chemical energy
to electric energy by a path that can avoid the thermodynamic limitation on
efficiency imposed by the Carnot relation - has intrigued scientists and
engineers for more than a century. In 1839-1842, Sir William Grove probably
invented the first fuel cell. He used platinum-catalyzed electrodes to combine
hydrogen and oxygen so as to produce electricity. It is interesting to note
that almost 123 years later the Gemini fuel cell used the same catalyst, though
probably in different physical form. One major trouble with Grove's cell was
that its voltage fell off badly when an appreciable current drain was put on it.
In 1889, Mond and Langer made a hydrogen-oxygen cell with perforated plati-

num-sheet electrodes, catalyzed by platinum black. This cell produced 1.46 watts at 0.73 volt at about 50 percent efficiency. However, it contained 1.3 grams of platinum and required pure hydrogen and oxygen. Thus, its capital cost made it a poor buy as an electric generator. In addition, to complicate things, it ran well only on pure hydrogen and oxygen.

SOURCE: "Hydrocarbon - air fuel cell systems" by C.G. Peattie. IEEE Spectrum (June 1966) p. 69.

SEE ALSO: "On voltaic series in combination of gases by platinum" by W.R. Grove Phil. Mag. Vol. 14. (1839) p. 127.

ALSO: "On a gaseous voltaic battery" by W.R. Grove. Phil. Mag. Vol. 21 (1842) p. 417.

NOTE: The principle of electrochemical fuel cells is by no means new, in fact the first cell was described in 1839 by Sir William Grove. This was a hydrogen fuel device with a sulphuric acid electrolyte and blacked platinum electrodes, generating approximately one volt at a very small current. The first power fuel cell was a 5 kilowatt unit demonstrated in 1959 by the English engineer, F.T. Bacon, and employed hydrogen-oxygen fuel with an alkaline electrolyte and sintered nickel electrodes. Rights to the development of this cell were obtained by Leesona-Moos Laboratories - a research subsidiary of Pratt & Whitney Aircraft Corp. - from N.R.D.C., and a modified version was used in the Apollo spacecraft.

SOURCE: "Fuel cells and their development in the U.K." by W.S.E. Mitchell Design Electronics (Feb. 1966) p. 34.

1839   MICROFILMING   Dancer (U.K.)

Shortly after the publication of Daguerre's invention of making photographs in 1839, Dancer produced in England photographs of documents of strongly reduced size (1:160) having a side length of abt. 3 mm. The knowledge of the possibility to produce reduced-size photographs prompted in 1853 the English astronomer John Herschel to suggest, to store documents of general concern (e.g. reference works) in a reduced form, provided the reduction does not involve any hazard for the original documents. The same idea was advanced at the beginning of this century by members of a Belgian library, in order to make old handwritings or prints accessible to many people. Unfortunately, this idea has never been materialized.

SOURCE: "A brief historical review on microfilming" by H. Scharffenberg and R. Wendel. Jena Review. No. 1. 1976. p. 4.

1839   PHOTOVOLTAIC EFFECT   E. Becquerel (France)

The photovoltaic effect was discovered by Edmond Becquerel as early as 1839. In 1873, Willoughby Smith first observed the photoconductivity of selenium. In 1887, finally, Heinrich Hertz described the photoemissive effect of ultraviolet light on metal electrodes.

SOURCE: "Beam-deflection and photo devices" by K. Schlesinger and E.G. Ramberg. Proc. IRE (May 1962) p. 991.

SEE ALSO: "On electric effects under the influence of solar radiation" by E. Becquerel. Compt. rend. acad. sci. Vol. 9. (1839) p. 561.

ALSO: "Effect of light on selenium during the passage of an electric current" by W. Smith. Amer. J. Sci. Vol. 5. (1873) p. 301.

1843   FACSIMILE REPRODUCTION   A. Bain (U. K.)

For the purpose of this review facsimile is considered to be a method by which printed, handwritten and graphic data may be transmitted over communication channels and received in the form of a hard copy. Its origin dates back to 1843, when the Scottish inventor Alexander Bain patented an "automatic electrochemical recording telegraph". Next came Frederick Bakewell's cylinder and screw arrangement on which many of the present-day facsimile systems are based.

In Europe facsimile equipment has been commercially available since 1946. In recent years the technology advances in electronics and the drastic fall in semiconductor prices have led to the replacement of bulky separate facsimile transmitters and receivers by small transceivers.

SOURCE: "Facsimile - a review" by J. Malster and M. J. Bowden. The Radio and Electronic Engineer. Vol. 46. No. 2. (Feb. 1976) p. 55.

1845   CIRCUITS - KIRCHOFF's LAWS   G. R. Kirchoff (Germany)

Two laws that express the behaviour of an electrical network. In 1845 he gave the laws for closed circuits, extending these to general networks (1847) and to solid conductors (1848).

1st law - The total current arriving at any point in an electric network must be zero.

2nd law - The sum of the electromotive forces around any closed circuit is equal to the sum of the IR drops around the circuit.

1847   MAGNETOSTRICTION   J. Joule (U. K.)

Magnetostriction is a wellknown pheomenon in which the mechanical dimension of a magnetic material is altered as the magnetization is varied; and in which, conversely, the magnetization is altered as the dimension is changed. Thus, an alternating current applied through a coil wrapped around a specimen can induce mechanical vibrations in it; and altrnately, mechanical vibrations set up in such a specimen can transform the mechanical energy into electrical energy in a coil wound around it. Electro-mechanical interactions were observed as early as 1847 with Joules discovery of magnetostriction.

SOURCE: "Solid State devices other than Semiconductors" by B. Lax and J. G. Mavroides. Proc. IRE (May 1962) p. 1014.

SEE ALSO: "On the effects of magnetism on the dimensions of iron and steel bars" J. P. Joule. Phil. Mag. Vol. 30. (April 1847) p. 226.

1850   THERMISTOR   M. Faraday (U.K.)

   The temperature-sensitive non-linear resistors are known generally as thermistors, a name coined by the Bell Telephone Laboratories (of the USA). They are, however, over 100 years old, for Faraday discovered that silver sulphide possessed a high negative temperature coefficient (although in this case the conduction is ionic and not electronic, and the material therefore suffers from polarization effects). Uranium oxide was used in Germany, but with this conduction is also ionic and operation is unstable. A magnesium titanate spinel was introduced in 1923, and in the U.S.A. about 1912, boron was found to possess negative temperature characteristics. From 1930 onwards the Bell Telephone Laboratories devoted many years of intensive research to the problem and showed that combined oxides of manganese and nickel had valuable properties. They also found that varying the ratio between the manganese and nickel varied these properties. The effects of adding small amounts of copper, cobalt and iron were also investigated.

   Today, these oxides, treated to become uniphase, are in general use, and are made into beads, rods, blocks, etc.

SOURCE: "Fixed Resistors" 2nd Edition. G. W. A. Dummer. Pitman. London (1967) p. 147.

1851   RELATION BETWEEN THEORY OF   Lord. Kelvin (U.K.)
         MAGNETISM & ELECTRICITY

   Introduced for the first time the vectors, later termed magnetic induction and magnetic force by Maxwell.

1852   THIN FILMS (Sputtering process)   W. R. Grove (U.K.)

   Although the use of cathodic sputtering as a method for the deposition of thin films predates vacuum evaporation by many years, the latter has received far more widespread application because evaporation is more convenient for many materials and generally gives higher deposition rates. In recent years, however, it has been found that certain materials are more conveniently deposited by sputtering. In some cases it is impossible to deposit materials by any other means.

SOURCE: "Thin-film Circuit Technology" by A. E. Lessor, L. I. Maissel and R. E. Thun. IEEE Spectrum (April 1964) p. 73.

SEE ALSO: W. R. Grove Phil. Trans. Roy. Soc. London. Series B. Vol. 162 (1852) p. 87.

1854   COMPUTERS (Calculating Machines)   P. G. Scheutz (Sweden)

   Pehr Georg Scheutz built a difference engine in Stockholm inspired by Babbage's ideas and displayed it in London in 1854 with considerable help from Babbage. The machine had four differences and fourteen places of figures and was capable of printing its own Tables (Scheutz was a printer).

SOURCE: "The computer from Pascal to von Neumann". by H. H. Goldstine.
Princeton Univ. Press 1972. p. 15. (Reprinted by permission)

1855   COLD CATHODE DISCHARGE TUBE   J. M. Gaugain (France)

        Experiments with low-pressure glow discharges started very early in
the electrical art. Most of them were in small diameter glow tubes, often in
the form called Geissler tubes. The first recognition of the fact that such a
glow tube having its two electrodes of different size was capable of rectifying
the oscillating current from an induction coil appears to have been that of
Gaugain in 1855. However, for many years, the only use for these discharges
was as light sources. The early glow lamps all required high voltage excitation.
With the availability of the rare gases neon and argon and by means of low work-
function cathodes, glow tubes were developed for low voltage applications.

SOURCE: "The development of gas discharge tubes" by J. D. Cobine. Proc. IRE
(May 1962) p. 972

SEE ALSO: J. M. Gaugain, Compt. rend. Acad. Sci. Paris. Vol. 40 (1855) p. 640.

1856   LOW PRESSURE DISCHARGE TUBES   H. Geissler (Germany)

        In 1856, Heinrich Geissler, an artist and skillful glass blower of Bonn,
Germany, originated the low pressure discharge tubes that were to bear his
name. The Geissler tubes were long, small-bore glass tubes, usually shortened
by the use of many coils and bends, which were filled with various gases at low
pressures and originally excited by high-voltage alternating current. Many
beautiful effects could be produced by Geissler tubes filled with different gases
and they were often used for decorations. As, for example, a display used to
commemorate Queen Victoria's Diamond Jubilee. However, sputtering of the
electrodes together with gas clean-up resulted in a short life for the tube. The
principal use was for spectral analysis and lecture demonstrations.

SOURCE: "The development of gas discharge tubes" by J. D. Cobine. Proc. IRE
(May 1962) p. 970.

SEE ALSO: "The electric lamp industry" A. A. Bright Jr., MacMillan Co.
New York. N. Y. p. 218 et seq. (1949)

ALSO: W. DeLaRue, H. W. Muller and W. Spottieswoode. Proc. Roy. Soc.
(London) Vol. 23. p. 356. (1875)

1857   MERCURY ARC LAMP   J. T. Way (U. K.)

        The first public demonstration of a mercury arc lamp was by Prof. J. T.
Way on the Hungerford Suspension Bridge in London on 3rd September, 1860. Two
British patents were issued to him in 1857. The electric arc was first used commercially for illumination in Paris in 1863. Much later, low-pressure arc
"tubes" were used for illumination. In 1879 John Rapieff described mercury arc
lamps in British Patent No 211 but there appears to be no evidence that they
were built. Peter Cooper Hewitt showed in public his mercury-arc lamp on
April 12, 1901. Georges Claude, a French inventor, demonstrated the first Neon
sign, an improvement of the Geissler tube, at the Grand Palais in Paris in 1910.
Developments in luminous tube discharges were made by Moore in 1920. Since

these tubes did not have a high light output, they were largely confined to sign applications.

SOURCE: "The development of gas discharge tubes" by J.D. Cobine. Proc. IRE (May 1962) p. 970.

SEE ALSO: British Patents issued 1857.

1860   BATTERY (Secondary)   Planté (France)

The secondary battery business dates back to Planté's discovery of the lead-acid system in 1860.

Secondary or storage cells are electrochemical cells which after discharge can be restored to their original chemical state by passing the current in the reverse direction. Although they have the same set of basic components as primary cells, the anodes and cathodes of secondary cells have a more stringent requirement, in that the electrode reactions have to be reversible. This requirement immediately limits the number of electrode materials available for secondary cells. At present, lead, cadmium, iron and zinc anode materials, and lead dioxide, nickel dioxide and silver oxide cathode materials are the only ones used in commercial secondary cells.

SOURCE: "Batteries" by C.K. Morehouse, R. Glicksman and G.S. Lozier. Proc. IRE (Aug. 1958) p. 1474/5.

1860   MICROPHONE (Diaphragm type)   J.P. Reis (Germany)

Earliest among microphone diaphragms - perhaps because of its similarity to the eardrum - was a stretched flat membrane (actually a sausage skin) used by Reis to actuate a loose metal-to-metal contact. A stretched flat membrane made of metal or very thin metallized plastic is used in present-day electrostatic microphones. This diaphragm is typically clamped at its periphery by a ring and stretched to any desired tension by a threaded ring.

SOURCE: "A century of microphones" by B.B. Bauer. Proc. IRE (May 1962) p. 720

SEE ALSO: "Ueber Telephone durch den galvanischen strom" Jahnesbericht d. Physikalischen Vereins zu Frankfurt am Main. Germany (1860-61) p. 57.

1860   ELECTRON   J.J. Thomson (U.K.)

By improving the vacuum of Hertz's cathode ray experiments, Thomson got deflections which, combined with the long-known deflections by a magnet, determined the ratio of the charge to the mass of the supposed particles. This ratio e/m was over 1,000 times larger than the ratio for hydrogen, the lightest atom known. Thomson considered that this was due to the smallness of the mass, and that particles with this small mass were universal constituents of matter, since they were the same whatever the chemical nature of the gas carrying the discharge and the electrodes through which it entered and left. He examined two other cases of the discharge of electricity, namely, those from a hot wire negatively charged and from a negatively charged zinc plate illuminated by ultra-violet light. In both he found charged particles with the same e/m ratio as the cathode rays, and in the second was able to measure the actual charge by condensing drops of water on the particles to form a mist, finding the size of the drops from their rate of fall. The results agreed with the supposed value of the charge on a hydrogen atom as far as this was then known. Thomson called these new light particles "corpuscles" but later adopted the work "electron", invented a few years before by G.J.Stoney for the charge on a hydrogen atom regarded as a natural unit of charge.

SOURCE: "A Biographical Dictionary of Scientists" T.I.Williams. Adam and Charles Black, London 1969. p.511.

SEE ALSO: "The discovery of the electron" by D.L.Anderson. Van Nostrand Reinhold. New York,1964.

1865   RADIO WAVE PROPAGATION   J.C.Maxwell (U.K.)

In his first paper on electromagnetism "On Faraday's Lines of Force" (1855-56) Maxwell set up partial analogies, between electric and magnetic lines of force and the lines of flow of an incompressible fluid. In a series of magnificent papers in 1861-2 he gave a fully developed model of electromagnetic phenomena viewed in the light of the field concept of Michael Faraday of whose validity Maxwell had become fully persuaded by 1858. Adopting the belief of William Thomson (Lord Kelvin) in the rotary nature of magnetism, a magnetic tube of induction was represented by a set of cells roating about the axis of the tube, interference between the rotations of neighbouring tubes being avoided by rows of intervening cells (in the manner of idle wheels) which corresponded to electric currents. By means of this model Maxwell was able to give an elegant qualitative interpretation of all the known phenomena of electromagnetism. By introducing the notion of elasticity he was then able to give a quantitative description of the propagation of a disturbance in the model. Reinterpreted in terms of the electromagnetic field, this implied that a disturbance in the electromagnetic field should travel with a speed equal to the ratio of the electrodynamic to the electrostatic units of electric force.

SOURCE: "A Biographical Dictionary of Scientists" T.I.Williams. Adam and Charles Black, London 1969. p.358.

SEE ALSO: "A Dynamical Theory of the Electromagnetic Field" J.C.Maxwell. Proc.Royal Soc.(London) Vol.13. pp.531-536. December 8th,1864.

1868   BATTERIES (Léclanché cell)   G.Leclanche   (France)

The Léclanché dry cell is perhaps the best known cell in common use today. It is widely used in flashlights and other such equipment. This type of cell was originally described by Georges Léclanché in 1868 and has undergone many improvements since that time. Basically, it consists of a nearly pure (99.99 per cent) zinc negative terminal, a carbon positive terminal, and a mixture of ammonium chloride, manganese dioxide, acetylene black, zinc chloride, chrome inhibitor and water. The mixture acts as a depolarizing agent to reduce the formation of hydrogen bubbles on the positive electrode as discharge takes place. Improvements which have been made include leakproofing, longer shelf life, pepped-up depolarizers, improved insulation and miniaturisation.

SOURCE: "Survey of electrochemical batteries" by N.D. Wheeler. Electro-
Technology (June 1963) p.68.

SEE ALSO: Léclanché G. Les Mondes Vol.16. (1868) p.532, also Comptes
Rendus, Vol.83 (1876) p.54.

1868   RHEOTOME (Waveform Plotter)   H. Lenz   (Germany)

      Heinrich Lenz (of Lenz's law) developed a segmented commutator or
rheotome which could be arranged to sample a periodic waveform at known
points of the cycle and thus feed a train of pulses, each corresponding to the
amplitude of the waveform at that point, to a slowly responding galvanometer.
By plotting the deflections against the time in the cycle at which they occurred,
the complete waveform could be reconstructed. In 1868, Bernstein used a
rheotome to chart the time course of the action potential in a nerve fibre.

      By 1876, the capillary electrometer of Marey and Lippman and the
string galvanometer of Einthoven were available with a sufficient sensitivity and
speed of response to record bioelectric events directly. However, preceding
Einthoven's studies of the electrocardiogram, Marchand in 1877 and Englemann
in 1878 were able to chart the electrocardiogram using a rheotome.

SOURCE: "From torpedo to telemetry" by D. W. Hill. Electronics & Power
27th November, 1975. p.1111.

SEE ALSO: "Des variations electriques, des muscles et due couer en
particulier etudiees au moyen de l'electrometre de M. Lippmann" E. J. Marey.
Comptes. Rendues Acad. des Sci. (1876) 82, pp. 975-977.

ALSO: "Bettrage zur kenntnis der reizwelle und contractionswelle de
herzmuskels" R. Marchand. Plugers Arch. f. d. ges. Physiol., (1877) 15, p. 511.

1870   BATTERY (Standard Clark Cell)   L. Clark   (U.K.)

      In about 1870 Latimer Clark, an English engineer and electrician, intro-
duced a new kind of voltaic cell consisting of a positive electrode of mercury
covered with a paste of mercurous sulphate and a negative electrode of zinc. The
electrolyte was a saturated solution of zinc sulphate. After many determinations
Latimer Clark assigned to the cell a mean value of 1·457 volt at 15·5°C. The
Clark cell suffered from a very high temperature coefficient of e. m. f.
(1200μV/°C). However, in spite of this, the Chicago International Electrical
Congress in 1891 adopted the Clark cell together with the silver coulometer in
definitions of the ampere and the volt.

SOURCE: "Standard Cells by Muirhead" Muirhead Technique Vol.18. No. 3.
(July 1964) p.19.

SEE ALSO: Clark Cell - Proc. Royal Soc.(1872.) Vol. XX p. 444.

ALSO: Clark Cell - Philos. Trans. Vol. CIXIV (1874) p.1.

NOTE: See also Weston Cell (1891.)

1874   CAPACITORS, MICA   M. Bauer (Germany)

Mica sheet as dielectric came into commercial capacitor manufacture only about 1914-1918 - very largely because not only could it stand up to the mechanical shocks of gunfire better than glass, but it also enabled the size of the capacitors to be reduced substantially for the same effective performance. The drive of war requirements pushed this development to the fore, although the use of mica as a capacitor dielectric had been "invented" more than 60 years earlier.

SOURCE: "Electrical Capacitors in our Everyday Life" by P. R. Coursey. ERA Journal No. 6. Jan. 1959. p. 10.

SEE ALSO: "Physical properties of mica" by M. Bauer. Z. deut. geol. Ge. Vol. 26 (1874) p. 137.

ALSO: "Mica condensers" by E. Bouty, Comptes Rendus. Vol. 110(1890) p. 486.

ALSO: "Capacity of mica condensers" by A. Zeleny. Phys. Rev. Vol. 22 (1906) p. 651.

1876   ROLLED PAPER CAPACITOR   D. G. Fitzgerald (U.K.)

It appears that the rolled paper capacitor was first covered by a patent filed in 1876 by Fitzgerald, who described:

"the construction of a condenser with layers of paper and conductor (usually tin-foil) alternately interleaved with each other on to a cylinder, and the impregnation of such condenser with paraffin wax after rolling."

SOURCE: "History, Present Status and Future Developments of Electronic Components" by P. S. Darnell. IRE Transactions on Component Parts. September, 1958. p. 124.

SEE ALSO: "Improvements in Electrical Condensers or Accumulators" by D. G. Fitzgerald. British Patent No: 3466/1876. September 2, 1876.

ALSO: "Paper Condensers" by X. Boucherot l'Eclairage Electrique, February 12, 1898.

ALSO: "The Manufacture of paper condensers" by G. F. Mansbridge. J. Inst. Elec. Engrs. Vol. 41 (May 1908) p. 535.

ALSO: "The capacity of paper condensers" by A. Zeleny and Andrewes. Phys. Rev. Vol. 27. (1908) p. 65.

1876   TELEPHONE   A. G. Bell (U.S.A.)

It has been a hundred years since a faint but momentous sound was made by Thomas Watson when he plucked a reed of a rudimentary transmitter. But that sound travelled over wire and was heard in another room by Alexander

Graham Bell, who happened to be holding a similar reed-and-diaphragm apparatus to his ear. This was the first telephone signal.

Later, on Feb.14.1876, Bell filed for the now famous patent on the apparatus that he and Watson had been working on - just three hours before one Elisha Gray filed a caveat with the Patent Office, declaring that he was working on a similar device but had not yet perfected it. Had the timing been different, we now might have a "Ma" Gray, instead of Ma Bell.

The first telephone patent was issued to Bell on March 7, 1876, three days before the historic moment when the first intelligible human voice was transmitted over the new telephone. Bell, after spilling acid over his clothing, had called out, "Mr. Watson, come here. I want you!" Next year the first commercial telephone went into service when a Boston banker leased two instruments, each consisting of a simple wooden box that contained both transmitter and receiver. The talker had to alternately talk and listen.

According to the records, the seed idea for the Bell System was planted when a merchant named Thomas Sanders made a verbal offer to Bell to finance the telegraphic experiments. They reached a tentative agreement, and shortly afterwards a lawyer named Gardiner G. Hubbard made Bell a similar offer. The three put into writing an agreement dated Feb. 27.1875. and later signed a deed of trust, dated July 9, 1877, forming the Bell Telephone Company, Gardiner G. Hubbard, Trustee.

SOURCE: "Electronics" December 11th, 1975. p. 91.

SEE ALSO: "Alexander Graham Bell and the invention of the telephone" by J.E.Flood. Electronics & Power (March 1976) p. 159.
Editor's Note: This issue of Electronics & Power is devoted to the early history of the telephone.

ALSO: "The marriage that almost was" by M.F. Wolff. IEEE Spectrum (February 1976) p. 41.

ALSO: "The telephone, its invention and development" by M. Woolley. Telecommunication Journals. Vol.43. 111/1976 p.175.

1877   PHONOGRAPH   T.A.Edison   (U.S.A.)
(Gramophone)

Alexander Graham Bells' invention of the telephone, in 1876, drew attention to the problems of the reproduction of speech. One of those attracted was Thomas Alva Edison, who was all the more interested in the study of sound because he was partially deaf. Sometime in the fall of 1877, Edison wrapped a sheet of tinfoil around a cylinder, set a needle in contact with it, turned a crank to rotate the cylinder, and into a mouthpiece attached to the needle he shouted the nursery rhyme that begins: "Mary had a little lamb". After making a few changes, he cranked the cylinder again, and from the horn of the instrument he heard a recognizable reproduction of his voice. Thus was born the first phonograph. The date of invention was later recollected by Edison as August 13th, 1877, but there is some question as to how precise that date is. We know, however, that the patent application was filed December 24th, 1877, and the patent was granted on February 19th, 1878.

SOURCE: "Disk Recording and Reproduction" by W.S. Bachman, B. B. Bauer and P. C. Goldmark. Proc. IRE (May 1962) p. 738.

SEE ALSO: US Patent No: 200, 521 (T. A. Edison)

1877   MICROPHONE (Carbon)   T. A. Edison   (U. S. A.)

      Among the earliest devices intended for converting vibration into electrical impulses was Reis' loose metal-contact transducer which is reported to have transmitted tones of different frequencies, but not intelligible speech. This latter event seems first to have been achieved by Bell, using a magnetic microphone, on June 3rd, 1875. However, Bell's microphone proved not to be sufficiently sensitive for telephone work, and the experiments of Berliner, Edison, Hughes and others soon thereafter introduced a long era of dominance for the loose-contact carbon transducer. To Edison goes the credit of being the first to design a transducer using granules of carbonized hard coal, still used in present-day microphones.

      The carbon granules are made of deep-black "anthraxylon" coal ground to pass a 60-80 mesh, treated chemically and roasted in several stages under a stream of hydrogen. This drives out volatile matter, washes out extraneous compounds, and carnonizes the coal. The last step of the process is magnetic and air-stream screening to eliminate iron-bearing and flat-shaped particles.

SOURCE: "A Century of Microphones" by B. B. Bauer. Proc. IRE (May 1962) p. 721.

SEE ALSO: T. A. Edison. U. S. Patent No: 474, 230 filed April 27th, 1877. Also U. S. Patent No: 474, 231/2.

1877   LOUDSPEAKER (Moving coil type) E. W. Siemens (Germany)

      The motor mechanism consisting of a circular coil located in a radial magnetic field was first disclosed by Siemens. Lodge, Pridham and Jenson and others contributed to the suspension system. However, there were very few developments in loudspeakers in the twenty-seven years following Lodge's disclosure. A breakthrough in the dynamic loudspeaker was made by Rice and Kellogg in 1925. The success of the development was due to their recognition of three physical factors with relation to the action and design of a direct radiator loudspeaker. The first is that the sound-power output of a loudspeaker is the product of the mechanical resistance due to sound radiation and the square of the velocity of the diaphragm. The second is that sound radiation from a small vibrating diaphragm gives rise to a mechanical resistance which is proportional to the square of the frequency. The third is a vibrating system which is mass controlled. It follows then that, if the fundamental resonance occurs below the lowest frequency of interest, the complementary variations of the second and third factors which control the sound output as given by the first factor conspire to provide a uniform response up to the frequency region at which the assumptions begin to fail. This was the contribution of Rice and Kellogg, and it continues to be the basic precept that guides the design of all direct-radiator loud-speakers.

SOURCE: "Loudspeakers" H. F. Olson. Proc. IRE. (May 1962.) p. 730.

SEE ALSO: "Electroacoustics" F. V. Hunt.   John Wily and Sons Inc. New York
N. Y. 1954.

ALSO: E. W. Siemens. German Patent No: 2355 filed December 14th, 1877.

ALSO: O. J. Lodge. British Patent No: 9712 filed April 27th, 1898.

ALSO: E. S. Pridham and P. L. Jenson. U. S. Patent No: 1, 448, 279. filed April
28th, 1920.

ALSO: "Notes on the development of a new type of hornless loudspeaker" by
C. W. Rice and E. W. Kellogg. Trans. AIEE, Vol. 44. pp. 461-475 April, 1925.

1878   CATHODE RAYS   Sir W. Crookes   (U. K.)

In his Bakerian lecture of 1878 and his British Association Lecture of
1879, he announced various striking properties of "molecular rays" including
the casting of shadows, the warming of obstacles and the deflection by a magnet.
The title "Radiant Matter" employed by Crookes in his British Association
Lecture of 1879 referred to ordinary matter in a new state in which the mean
free path was so large that collisions between molecules could be ignored.

SOURCE: "A Biographical Dictionary of Scientists" T. I. Williams. Adam and
Charles Black, London 1969. p. 120.

1878   CARBON FILAMENT   J. W. Swan, C. H. Stearn, F. Topham
        INCANDESCENT LAMP   and C. F. Cross   (U. K.)

Swan invented a carbon filament incandescent lamp and a squirting
process to make nitro-cellulose fibres for lamp filaments. Stearn was an
expert in the production of high vacua and Topham the glass-blower who made
the glass globes. Cross discovered the viscose process for the fibres.

SOURCE: "The Sources of Invention" by J. Jewkes, D. Sawers and
R. Stellerman. MacMillan, London 1958. p. 59.

1879   HALL EFFECT   E. H. Hall (U. K.)
Hall voltage, linear function of magnetic flux.

If a current of particles bearing charges of a single sign and constrain-
ed to move in a given direction is subjected to a transverse magnetic field, a
potential gradient will exist in a direction perpendicular to both the current and
the magnetic field.

SOURCE: The Encyclopaedia of Physics. 2nd Ed. Editor R. M. Besancon.
Van Nostrand, New York 1974. p. 400.

1880   PIEZO ELECTRICITY   J. Curie and P. Curie   (France)

The relation between voltage generated and mechanical pressure on
crystallographic materials.

SOURCE: "Developpement, par pression, de l'electricite polaire dans les
cristaux hemiedres a faces inclinees" J. Curie and P. Curie. Comp. Rend.
Vol. 91. pp. 294-295. July-December, 1880.

1882   WIMSHURST MACHINE   J. Wimshurst (U.K.)

The Wimshurst machine generates electrostatic electricity by friction. It consists of two discs which revolve in opposite directions. The discs are made of an insulating material such as glass and near the peripheries of the discs are mounted small sections of sheet conductor. The charges are generated on the conductor sections by the action of brushes which graze the sectors as they rotate and are picked up by combs. These charges are stored in capacitors and are used to produce a spark between a pair of ball conductors.

REFERENCE: "Text Book of Physics" by J. Duncan and S. C. Starling. Part 5 - Magnetism and Electricity. MacMillan & Co. London (1926) p. 960.

REFERENCE: "Encyclopedic Dictionary of Electronics and Nuclear Engineering" by R. I. Sarbacher. Pitman, London (1959) p. 1402.

1884   BATTERY (Zinc-mercuric oxide cell)   C. L. Clarke (U.S.A.)

The alkaline zinc-mercuric oxide system was first suggested by Clarke in 1884. Although there were a number of additional attempts made over the years to design a practical cell using this system, it was not until early in World War II that a commercially usable mercuric oxide dry cell was invented by Ruben.

SOURCE: "Batteries" by C. K. Morehouse, R. Glicksman and G. S. Lozier. Proc. IRE (Aug. 1958) p. 1467.

SEE ALSO: C. L. Clarke. U. S. Patent No: 298, 175. (May 6, 1884)

ALSO: "Balanced alkaline dry cells" by S. Ruben. Trans. Electrochem. Soc. Vol. 92 (1947) p. 183.

1884   SWITCH, QUICK BREAK   J. H. Holmes (U.K.)

The loose-handle quick-break switch was invented by J. H. Holmes in 1884. For this device he was granted British Patent No: 3256 of 1884, under the title: "Improvements in or applicable to switches or circuit closers for electrical conducting apparatus ".

SOURCE: "Patents for Engineers" by L. H. A. Carr and J. C. Wood. Chapman & Hall, London 1959. p. 95.

1885   RESISTOR (Moulded carbon composition type)   C. S. Bradley (U.K.)

It is of passing interest to note that the earliest moulded-rod composition-type resistor of which the author has been able to trace any record dates back before the days of radio. In 1885, a moulded-composition resistor was patented, comprising a mixture of carbon and rubber heated and moulded to shape and subsequently vulcanized to a hard body.

SOURCE: "Fixed resistors for use in communication equipment", by P. R. Coursey. Proc. IEE Vol. 96 Pt. III (1949) p. 169.

SEE ALSO: C. S. Bradley   British   Patent No: 8076/1885.

<u>ALSO</u>:   M. Slattery   U. S. Patent No: 354, 275   (1885)

<u>ALSO</u>:   D. C. Voss.   U. S. Patent No: 573, 558   (1896)

1885   <u>TRANSFORMER (Distribution)</u>   M. Deri   (Austria)

For the earliest patent covering the use of the transformer in a distribution system it is necessary to refer to German Patent No: 33951 of 1885, since no corresponding patent was applied for in the United Kingdom.

This application was made on 18th February 1885 by Max Deri (in this case described as being of Vienna) under the title (translated) of "Improvements in the Distribution of Electricity".

<u>SOURCE</u>:   "Patents for Engineers" by L. H. A. Carr and J. C. Wood. Chapman and Hall. London 1959. p. 91.

1885   <u>TRANSFORMER (Power)</u>   C. Zipernowski, M. Deri & O. T. Blathy (Hungary)

The earliest patent covering the construction of the transformer appears to be that of Carl Zipernowski, Max Deri and Otto Titus Blathy, all of Budapest, and since this patent was applied for and granted in this country, the British version can be quoted here.

The date of application was 27th April, 1885, the patent being numbered 5201 in that year under the title: "Improvements in Induction Apparatus for Transforming Electric Currents".

<u>SOURCE</u>:   "Patents for Engineers" by L. H. A. Carr and J. C. Wood. Chapman and Hall. London 1959. pp. 89/90

1887   <u>GRAMOPHONE</u>   E. Berliner (U. S. A.)
(Phonograph)

On May 4th, 1887, Emile Berliner applied for a patent on what he called a "Gramophone" to distinguish it from Edison's phonograph of ten years earlier and from Bell's and Tainter's Graphophone of one year earlier. The first figure of the drawings in Berliner's patent shows a record wound on a cylindrical support but by 1888, when he introduced his first model, he had changed to a flat-disk record. The groove in this record had a lateral side-to-side movement, as against the vertical "hill and dale" system which had been employed by others. This lateral recording process was reminiscent of Leon Scott's phonoantograph, which used a diaphragm and hog bristle to trace a record of sound vibrations on lamp-blacked paper some thirty years earlier.

Berliner also used lamp-black as the recording medium, and combined this method with an etching process which permitted transfer of the original engraving to copper or nickel. Thus Berliner achieved a permanent master recording, and for the first time mass duplication of records was possible. No longer did artists have to repeat each number endless times.

By 1895, Berliner had developed a system utilizing many ideas of his own and others: Scott's lateral groove, his own flat disk, and a coating of Bell's and Tainter's wax. The system stood up as the industry standard for half a century, thus Berliner deserves a mantle as the father of disk recording and reproduction.

SOURCE: "Disk Recording and Reproduction" by W.S. Bachman, B.B. Bauer and P.C. Goldmark. Proc. IRE (May 1962) p. 738-739

SEE ALSO: "The gramophone and the mechanical recording and reproduction of musical sounds" by L.N. Reddie. J. Roy. Soc. Arts. Vol: LVI, pp. 633-649 May 8th, 1908.

1887   AERIALS (Radio Wave Propagation)   H.R. Hertz (Germany)

It remained for Heinrich Hertz to prove the existence of electric waves in space as predicted by Maxwell. The first true antenna appears to have been used by Hertz in his classical experiments at Karlsruhe in 1887. His antenna consisted of two flat metallic plates, 40 cm square, each attached to a rod 30 cm long. The two rods were placed in the same straight line, and were provided at their nearer ends with balls separated by a spark gap about 7-mm long. The spark gap was energized by a Ruhmkorff coil. In order to detect the radiated waves, Hertz employed a receiving circuit consisting of a circular loop of wire broken by a microscopic gap. The radius of the loop was 35 cm, which was found by experiment to be the proper size to be in resonance with the oscillator.

SOURCE: "Early history of the antennas and propagation field until the end of World War I, Part I - Antennas" by P.S. Carter and H.H. Beverage. Proc. IRE (May 1962) p. 680)

SEE ALSO: "Ueber sehr schnelle elektrische Schwingungen" H. Hertz. Ann. Physik und Chemie (Wiedeman) N.F. Vol. 31. pp. 421-448. May 15, 1887.

1887   SWITCH, QUICK MAKE & BREAK   J.H. Holmes (U.K.)

Following Holmes invention of the quick-break switch (see 1884) a later variant provided for a loose-handle operating a spring over a dead centre by means of a toggle action, so that both quick-make and quick-break were obtained. (British Patent No: 5648 of 1887).

SOURCE: "Patents for Engineers" by L.H.A. Carr and J.C. Wood. Chapman & Hall. London 1959. p. 95.

1888   INDUCTION MOTOR   N. Tesla (U.S.A.)

Tesla's "master" patent covering the polyphase induction motor was taken out, inter alia, in Great Britain; the British Patent, No: 6481 of 1888, may therefore be used as a reference.

Its title was "Improvements relating to the Electrical Transmission of Power and to Apparatus therefor", and it was granted in the name of Nikola Tesla of the City and State of New York, U.S.A.

The specification is long and detailed with 18 diagrammatic figures.

SOURCE: "Patents for Engineers" by L.H.A. Carr and J.C. Wood. Chapman and Hall, London 1959. pp. 96/7

1889   COMPUTERS (Tabulating Machines)   H. Hollerith   (U.S.A.)

Hollerith worked on a machine for tabulating population statistics for the 1890 Census in the U.S.A. which he patented in 1889. He used a system of holes in a punch card to represent various characteristics: such as male or female, black or white, age, etc. The cards were 6 5/8 by 3 $1\frac{1}{4}$ inches in size, which he chose because it was the size of a dollar bill. Each card contained 288 locations at which holes could be made.

SOURCE: "The computer from Pascal to von Neumann" by H. H. Goldstine. Princeton Univ. Press 1972. p. 65. (Reprinted by permission)

SEE ALSO: "An electric tabulating system" by H. Hollerith, reprinted in 'The Origins of Digital Computers" edited by B. Randell, Springer-Verlag, Berlin (1973) p. 129.

1890   COHERER   E. Branley (France)

In Paris, Edouard Branly, physics professor at the Catholic University observed in 1890 that metal filings, when subjected to "Hertzian waves" behaved very strangely. Normally, filings do not transmit an electric current because there are air spaces between them; but when placed with the range of electro-magnetic waves, the filings fuse a little together, enough to offer a conducting path to an electric current. The filings remain a conductor until they are disturbed by shaking or tapping.

Branly called the little glass tube in which he placed his filings "coherer"; it was the first form of a "detector" for electro-magnetic waves.

SOURCE: "A History of Invention" by E. Larsen. J. M. Dent & Sons, London and Roy Publishers, New York, 1971. p. 278.

1891   BATTERY (Standard Weston cell)   E. Weston (U.S.A.)

In 1891, Dr. Edward Weston, an Anglo-American from New Jersey, filed a patent which was granted in 1893 disclosing a cell in which the electrolyte of the Clark cell was replaced by a saturated solution of cadmium sulphate and the zinc negative electrode was replaced by cadmium amalgam, the depolariser being mercurous sulphate as before. This cell was a definite improvement on the Clark cell since the temperature coefficient of e.m.f. was only about $-40\mu V/^{\circ}C$. Considerable improvements were made by various workers from 1893 onwards and finally the London Conference held in 1908 authorised the appointment of a special international committee. As a result of the work of this committee, Weston's cadmium cell was assigned the value of 1·01830 International volts at 20°C.

SOURCE: "Standard cells by Muirhead" Muirhead Technique. Vol. 18. No. 3. (July 1964) p. 19.

SEE ALSO: Weston cell - U.S.A. Patent No: 494, 827 (1893) also British Patent 640, 812797, 381.

1893   WAVEGUIDES (J.J.Thomson (U.K.)

Perhaps the first analysis suggesting the possibility of waves in hollow
pipes appeared in 1893 in the book "Recent Researches in Electricity and
Magnetism" by J.J.Thomson. This book, which was written as a sequel to
Maxwell's "Treatise on Electricity and Magnetism" examined mathematically
the hypothetical question of what might result if an electric charge should be
released on the interior wall of a closed metal cylinder. Even now, this problem is of considerable interest in connection with resonance in hollow metal
chambers. A much more significant analysis, relating particularly to propagation through dielectrically filled pipes, both of circular and rectangular cross
section, was published in 1897 by Lord Rayleigh.

SOURCE: "Survey and History of the Progress of the Microwave Arts"
by G.C.Southworth. Proc.IRE (May 1962) p.1199.

SEE ALSO: "Recent researches in electricity and magnetism" by J.J.Thomson
(1893) p.344.

ALSO: "On the passage of electric waves through tubes or the vibrations of
dielectric cylinders" Lord Rayleigh. Phil.Mag.Vol_43 (Feb.1897) p.125.

1895   X-RAYS   W.K.Rontgen (Germany)

On 8th November 1895, while experimenting with a Crookes's tube
(see W.Crookes) covered with an opaque shield of black cardboard, Rontgen
noticed that, when a current passed through the tube, a nearby piece of paper
painted with barium platinocyanide fluoresced. In a series of classical papers
(1895-7) he described the properties of the new, so-called X-rays, but his
attempts to detect their interference by crystals were unsuccessful.

SOURCE: "A Biographical Dictionary of Scientists" T.I.Williams. Adam and
Charles Black, London 1969. p.448.

1896   WIRELESS TELEGRAPHY   G. Marconi   (Italy)

On 2nd June 1896, Marconi took out in the United Kingdom the first
patent for wireless telegraphy based on Hertz's discoveries, though exploiting
radiations of a much longer wavelength. His apparatus consisted of a tube-like
receiver or "coherer" connected to an earth and an elevated aerial; its signals
were at first transmitted over one hundred yards, a satisfactory demonstration
being arranged from the roof of the London General Post Office. Ship to shore
communication was established in the following year, when Marconi formed a
Wireless Telegraph Company in London for the exploitation of his patents in all
countries except Italy; this later developed world-wide affiliations. His first
transatlantic signals were made on 12th December,1901 from Poldhu in Cornwall
to St.John's, Newfoundland, where they were received through an aerial suspended from a kite.

SOURCE: "A Biographical Dictionary of Scientists" T.I.Williams. Adam and
Charles Black, London 1969. p.352.

1897 <u>RESISTOR (Carbon film type)</u>   T.E. Gambrell
                                       a n d   (U.K.)
                                       A.F. Harris

The earliest type of carbon-film resistor was also used many years before broadcasting and it was apparently forgotten when the greater demand arose. Some of these carbon-film resistors, notably those formed by spraying or otherwise applying the conducting coating and then baking on to a glass filament, were enclosed inside a glass tube with metal end-caps sealed on for terminal connections and with terminals connected to the filament coating by casting into a metal such as type-metal.

<u>SOURCE</u>: "Fixed resistors for use in communication equipment" by P.R. Coursey. Proc. IEE Vol. 96 Pt. III (1949) p. 170.

<u>SEE ALSO</u>: T.E. Gambrell and A.F. Harris. British Patent No: 25412/1897.

1897 <u>CATHODE RAY OSCILLOGRAPH</u>   F. Braun (Germany)

The cathode-ray oscilloscope for the study of the time variation of electron currents was developed by Ferdinand Braun in 1897, the same year in which J.J. Thomson measured the specific charge of the electron by its deflecti on in electric and magnetic fields. Ferdinand Braun constructed the first cathode-ray oscillograph at the University of Strassbourg in 1897.

Just like the early X-ray tube, the "Braun tube" used gas discharge phenomena for the emission and formation of an electron beam. Even after the introduction of a thermionic cathode into cathode-ray tubes by Wehnelt in 1905 an argon atmosphere of about 10^{-3} mm Hg was still retained in commercial oscilloscopes for another 25 years. The effect of ion focusing facilitated the formation of long, filamentary, electron beams.

<u>SOURCE</u>: "Beam-deflection and photo devices" by K. Schlesinger and E.G. Ramberg. Proc. IRE (May 1962) p. 991 and 995.

<u>SEE ALSO</u>: "On a method for the demonstration and study of currents varying with time". F. Braun.. Wiedemann's Ann. Vol. 60. (1897) p. 552.

<u>ALSO</u>: "Cathode Rays" J.J. Thomson. Phil. Mag. Vol. 4 (1897) p. 293.

1898 <u>MAGNETIC RECORDING</u>   Valdemar Poulsen (Denmark)

No one really knows how the idea of magnetic recording occurred to Poulsen. During his experiments he found that:

> "it would be possible to magnetize a wire to different degrees so close together that sound could be recorded on it by running the current from a microphone through an electromagnet and by either drawing the wire rapidly past the electromagnet or drawing the electromagnet rapidly past the wire."

This invention had several things to commend it: the wire or tape could be used over and over again by de-magnetizing it and recordings could be played thousands of times without destroying the quality. Poulsen invented this Telegraphone in 1898; with it he won the Grand Prix at the Paris Exposition in 1900. He filed an application for a Danish patent in 1898 and within two years he had filed additional patent applications in the United States and most European countries. These early patents suggested, as recording media, steel wires and tapes and discs of material coated with magnetisable metallic dust, though he himself used only steel wire and tape in his machinery. The basic principles enunciated by Poulsen are still applied in all types of modern magnetic recorders.

SOURCE: "The Sources of Invention" by J. Jewkes, D. Sawers and R. Stillerman. MacMillan & Co. London 1958. p. 326.

SEE ALSO: "The Development of the Magnetic Tape Recorder" Engineer. March 18th, 1949.

1900   BATTERY (Nickel-iron cell)   T. A. Edison (U.S.A.)

The nickel-iron-alkaline batteries as they exist today are essentially the same as discovered by Edison around 1900 and marketed in 1908. The negative electrode consists of pockets of active material grouped together to form a plate, while the positive plate is an assembly of perforated nickel tubes filled with nickel hydroxide and nickel.

SOURCE: "Batteries" by C. K. Morehouse, R. Glicksman and G. S. Lozier. Proc. IRE (August 1958) p. 1478.

SEE ALSO: "The Edison nickel-iron-alkaline cell" by F. C. Anderson, J. Electrochem. Soc. Vol. 99. (1952) p. 244C.

1900   BATTERY (Nickel-cadmium cell)   Junger & Berg   (Sweden)

The nickel-cadmium battery discovered by Junger and Berg about 1900 is closely related to the nickel-iron Edison battery. Both are mechanically rugged and will withstand electrochemical abuse in that they can be overcharged, ovdrdischarged, or stand idle in a discharged condition. The nickel-cadmium battery differs from the Edison battery in the use of cadmium anodes in place of iron.

SOURCE: "Batteries" by C. K. Morehouse, R. Glicksman and G. S. Lozier. Proc. IRE (August 1958) p. 1478.

SEE ALSO: "Storage Batteries" by G. W. Vinal. J. Wiley New York 4th Ed. (1955).

1900   CAPACITORS, CERAMIC   L. Lombardi (Italy)

Ceramic materials have been used for many years as electrical insulators. They are able to withstand severe working conditions because they are vitrified by firing at temperatures of the order of 1,200°C. Being completely inert they will withstand their rated working voltage indefinitely and retain their shape and physical characteristics under normal conditions.

SOURCE: "Fixed Capacitors" (2nd Edition) by G. W. A. Dummer. Pitman, London (1956) p. 115.

SEE ALSO: "An improved process for manufacturing thin homogenous plates, more particularly applicable for use in electrical condensers" L. Lombardi, British Patent No 9133 filed May 17, 1900.

ALSO: "Permittivity of titania" by W. Schmidt. Ann. Phys. Lpz. 4, 9(1902) p. 959. "Dielectric losses and breakdown strength of porcelain" by F. Beldi. Brown Boveri Rev. 18 (May 1931) p. 172.
"Insulating materials of the steatite group" by E. Schonberg. Elektrotech. Z. 54. (June 1933) p. 545.
 Tubular metallized ceramic capacitors. Porzellanfabrikkahla, Germany. Brit. Pat. No:440951/1934.

1901   FLUORESCENT LAMP   P. Cooper-Hewitt   (U.S.A.)

      The first low-pressure mercury discharge lamp was introduced at the beginning of the twentieth century by Peter Cooper-Hewitt, the American individual inventor. Sir Humphrey Davy had discovered the effect of a discharge through mercury early in the nineteenth century. Cooper-Hewitt's lamp was inefficient by modern standards, though better than contemporary incandescent lamps; it also produced the characteristic blue light of the mercury discharge lamp. It differed from the modern fluorescent lamp in being designed to produce visible radiation, not ultra-violet. In 1901 he used rhodamine dye, which fluoresces red, to improve the light's colour, but the rhodamine deteriorated too rapidly for this to be a success. The Moore and neon discharge lamps, introduced at much the same time as the Cooper-Hewitt, also contributed to the development of the fluorescent lamp: D. McFarlan Moore, an American individual inventor, was the first to apply the hot cathode used on it, and he also anticipated Wehnelt in constructing lasting electrodes. Georges Claude's introduction of the neon tube, and the desire to modify its colour, stimulated interest in fluorescent powders and in means of employing them with a lamp.

SOURCE: "The Sources of Invention" by J. Jewkes, D. Sawers and R. Stellerman. MacMillan, London 1958. pp. 298/9

SEE ALSO: "Lighting by Luminescence" by A. Claude. Light and Lighting. June 3, 1939.

1902   INDUCTION MOTOR   E. Danielson (Sweden)
        (SYNCHRONOUS)

      Danielson's synchronous induction motor is an excellent example of the sudden appearance of an invention which was complete at its first inception.
      It was not patented in Great Britain, but the full information is available in United States Patent No:694092 of 25th February, 1902, granted to Ernst Danielson, of Westeras, Sweden.
      After relating the advantages of the over-magnetised synchronous motors for eliminating lag (of current) and referring to the difficulty of starting such machines, the specification proceeds:

      "The invention consists, briefly, in combination with an ordinary induction-motor and a suitable resistance for connecting to the secondary element of said motor, a source of continuous electric currents (and) a switch arrangement so connected that the secondary part of the motor may by means of said switch arrangement either be

connected to the said resistance or to the said source of continuous
currents."

With reference to the self-synchronizing action, the specification states:-

"When the exciting current is supplied, the motor is changed from an
asynchronous to a synchronous one, provided, however, that the
exciting-current is strong enough to pull the motor in step."

The drawing and diagram of connections (Showing a three-phase-secondary
winding) which are attached to the specification are completely up-to-date, and
were it not for the somewhat archaic outlines of the machines, might have been
taken from a present-day text book or manufacturer's pamphlet.

SOURCE: "Patents for Engineers" by L.H.A. Carr and J.C. Wood. Chapman
and Hall, London 1959. pp. 97/8

1904   CAPACITORS, GLASS (TUBULAR)   I. Moscicki (U.K.)

The glass dielectric capacitor was later manufactured in tubular form-
known as the Moscicki tube - which provided the only form of capacitor available
for Marconi's early experiments in practical wireless communication; and in a
modified form, using flat glass plates interleaved with zinc sheets and all
immersed in oil, continued to provide the condensers for the spark wireless
transmitting apparatus up to and including part at least of the 1914-18 war
period.

SOURCE: "Electrical Capacitors in our Everyday Life" by P.R. Coursey.
ERA Journal. No.6. Jan.1959. p.10.

SEE ALSO: I. Moscicki "Improvements in electric condensers" British Patent
No:1307, filed January 18th 1904.

1904   TWO ELECTRODE TUBE   J.A. Fleming (U.K.)

Fleming received his early education in London, where he attended
London University; later he spent four years at Cambridge, and he was
appointed "electrician" to the Edison Electric Light Company in 1882. During
a visit to the United States in 1884 he visited Edison to discuss electric
lighting problems, and it is of particular moment to the story of the valve that,
during the visit, Edison demonstrated a discovery he had made a year before-
the Edison effect. Using a carbon-filament lamp in which a metal plate had
been sealed, Edison found that when the plate was connected through a
galvanometer to the positive terminal of the filament a current flowed, but no
current flowed when connection was made to the negative terminal. He used
the device to regulate the supply voltage in power stations. Fleming was very
interested in the phenomenon, and, on his return home, carried out researches in which he showed that "the space between the filament and the
metal plate is a one-way street for electricity".

It was on the 16th November,1904, that Professor J.A. Fleming
(1849-1945) described in British Patent Specification No:24850 a two-electrode
valve for the rectification of high-frequency alternating currents.

SOURCE: "Fleming and de Forest - an appreciation" by Captain C.F. Booth.
IEE pub. Thermionic Valves 1904-1954. IEE London 1955. p.1.

1906   <u>RADIO BROADCASTING</u>   R. Fessenden (U.S.A.)

   The first documented successful broadcasting of speech and music was conducted by Dr. Reginald Fessenden at Brant Rock, Mass., on Christmas eve, 1906, utilizing a 50 kc radio-frequency alternator which produced about 1 kw of power and which was built by the General Electric Co., under the direction of Dr. E. F. W. Alexanderson. Modulation was accomplished by means of a microphone which is believed to have been water-cooled and which was connected in the antenna circuit. Clear reception was obtained at many locations including ships at sea.

   Subsequently, Dr. Lee de Forest conducted experimental broadcasting in 1907 from his laboratory in New York City, in 1908 from the Eiffel Tower in Paris, and in 1910 from the Metro-olitan Opera House in New York City. These experiments, which were conducted with arc transmitters of about 500 watts power, modulated by microphones in the antenna-ground system, while successful, were handicapped by the high noise level inherent in arc transmitters.

<u>SOURCE:</u> "AM and FM Broadcasting" by R. F. Guy. Proc. IRE (May 1962) p. 811

<u>SEE ALSO:</u> "History of radio to 1926" by G. L. Archer. The American Historical Society, New York. N. Y. 1938.

1906   <u>THREE ELECTRODE TUBE</u>   L. de Forest (U.S.A.)

   While Fleming was developing his two-electrode valve, Dr. Lee de Forest was working in the United States on womewhat similar lines, and on the 25th October, 1906, de Forest applied for a patent for a three-electrode valve - a triode - as a device for amplifying feeble electric currents, the amplification being achieved by using a voltage on the intermediate electrode (grid) to control the plate current. A few months later de Forest extended the patent to cover the use of the valve as a detector. The introduction of the third electrode to provide an amplifier as compared with the two-electrode rectifier very greatly extended the potential applications of the thermionic valve, and much credit is due to de Forest for his achievement.

   Unfortunately, the invention of the triode led to considerable bitterness and litigation involving Fleming and de Forest, the former insisting to the end of his long life that de Forest's work was dependent on his own two-electrode valve. On the other hand, de Forest has always maintained that he was not aware of Fleming's patent before taking out his own. Initially the American courts held that de Forest's addition of the grid was dependent on Fleming's work. The story of the patent litigation did not end until 1943, when the United States Supreme Court decided that the original Fleming patent had always been invalid.

<u>SOURCE:</u> "Fleming and de Forest - an appreciation" by Captain C. F. Booth. IEE Pub. Thermionic Valves 1904-1954. IEE London 1955. p. 2.

1908   <u>GEIGER COUNTER</u>   E. Rutherford and H. Geiger (U.K.)

   About this time, experiments were being conducted with the collection of current by a positive wire and negative cylinder arrangement that were to have a profound effect on science. These experiments were reported in a paper by

Rutherford and Geiger, which showed that the number of charges of an ionizing event could be multiplied several thousand times by the ionizing action of electrons in the high field region near the wire. This was the start of what for a time were called Geiger-Muller counter tubes and now simply Geiger counter tubes. The technique of the proportional type of counter was established in 1928, and the following year, schemes for determining the coincidence of ionizing events were presented. Thus direction, scattering, absorption, etc., types of experiments were possible and the modern era of cosmic-ray and nuclear research developed.

SOURCE: "The development of gas discharge tubes" by J.D. Cobine. Proc. IRE (May 1962) p.971.

SEE ALSO: E. Rütherford and H. Geiger. Proc. Royal Soc. London A. Vol. 81. (1908) p.612.

1908   ELECTRONIC ORGAN   T. Cahill (U.S.A.)

Beginning with the 1900s there were many attempts to offer electric or electronic substitutes for organs, but none enjoyed any degree of commerical success. They were based on photo-optics, magnetic or electrostatic pre-recordings, vacuum tube or neon lamp oscillators, or amplified blown reeds, etc. One of these, called the Telharmonium, was invented by Thadius Cahill and demonstrated in 1908. The size of a small power-generating station, it consisted of almost a hundred alternator generators for all the frequencies of the scale. Then through a console of switches, synthesised musical signals were transmitted over telephone lines without benefit of amplifiers.

In 1935 Mr. Laurens Hammond, based on his synchronous electric clock, invented the first commercially successful mass-produced electric organ that started an industry. Since then, many manufacturers all over the world have joined this industry, offering instruments in a variety of sizes and prices that have transformed modern music.

SOURCE: "Electronics Engineer's Reference Book" Newnes-Butterworth London (1976) Chap.17. p.17 - 2

SEE ALSO: T. Cahill U.S. Patent No:1295691 (25th Feb.1919)

ALSO: L. Hammon U.S. Patent No:1956350 (24th April, 1934)

1908   TELEVISION (Electronic) A. A. Campbell-Swinton (U.K.)

Boris Rosing, of the St. Petersburg Technological Institute, seems to have been the first physicist who thought of using Braun's tube for the reception of images. As early as 1907 he suggested a system of remote electric vision, with a Nipkow disc for scanning the scene to be transmitted and a cathode-ray tube as the receiver. At about the same time the English inventor, A. A. Campbell-Swinton, also proposed a system of electronic television, but with cathode-ray tubes for transmission as well as for reception. He published his ideas in the scientific magazine NATURE in 1908, and elaborated them again in 1911 and 1920, explaining that the image transmitted in this way could be split up into, and reassembled from, about 400,000 points of different light value within $\frac{1}{25}$ of a second.

SOURCE: "A History of Invention" by E. Larsen. J. M. Dent & Sons, London and Roy Publishers, New York, 1971. p. 323.

1909 FERRITES (H.F.)   G. Hilpert   (Germany)
                           J. L. Snoek   (Holland)

The first proposal for high-frequency application was made in Germany in 1909 by Hilpert, who first synthesized such ferrites. However, the practical development and exploration of these materials did not come about until Snoek from the Philips Laboratories in Holland carried out extensive investigations on the high-frequency properties of such materials as manganese and nickel ferrite well into the UHF region.

SOURCE: "Solid-State devices other than semiconductors" by B. Lax and J. G. Mavroides. Proc. IRE (May 1962) p. 1012

SEE ALSO: "Genetische und konstitutive Zusammenhange in den magnetischen Eigenschaften bei Ferriten und Eisenoxyden" G. Hilpert. Berichte deutsch chemisch, Gesell. Vol. 42. p. 2248-2261. (1909)

ALSO: "New developments in ferromagnetic materials" J. L. Snoek. Elsevier Pub. Co. New York 1947.

1910 NEON LAMP   G. Claude   (France)

The majority of cold cathode tubes use a gas filling of which neon is the principal constituent. Development of the modern cold cathode tube may thus be said to date from 1898, the year in which Sir William Ramsay discovered neon. Ten years later Georges Claude began to isolate substantial quantities of a helium-neon mixture and in 1910 Claude exhibited two 38-ft. neon tubes. These were the precursors of the familiar neon sign and decorative lighting tube of today, in which the light comes principally from the long positive column.

Filament lamps and electric power were both expensive at that time. There was accordingly a strong incentive to develop a cheap and robust lamp of low power consumption and suitable for use on domestic supply voltages. Professor H. E. Watson has described the work which led to the appearance of the domestic neon lamp, first in Germany in 1918 and later in Holland. Once sputtering had been substantially reduced, the design of a 'beehive' neon for 220 V supplies presented few problems. For 110 V supplies, however, an alloy or activated cathode was required in order sufficiently to reduce the cathode fall. The successful development of such a tube by Philips in Holland stimulated renewed activity in the U. S. A. and eventually, in 1929, the General Electric Co., produced a miniature neon indicator.

SOURCE: "A survey of cold cathode discharge tubes" by D. M. Neale The Radio and Electronic Engineer. February 1964. p. 87.

SEE ALSO: "The development of the neon glow lamp (1911-61)" by H. E. Watson Nature 191, No. 4793. p. 1040-1 9th September, 1961.

1912   CIRCUITRY (REGENERATIVE)   de Forest, Armstrong E.H.
                                          and Langmuir I.   (U.S.A.)
                                          Meissner   (Germany)

The regenerative circuit was invented in 1912 by de Forest, Armstrong and Langmuir in the United States and by Meissner in Germany. After 20 years of litigation, the United States Supreme Court finally decided in de Forest's favour.

In the regenerative detector circuit RF energy is fed back from the anode circuit to the grid circuit to give positive feedback at the carrier frequency, thereby increasing the sensitivity of the circuit.

Regenerative receivers marked a big step forward in providing greatly increased sensitivity. Inherently they provided large amplification of small signals and small amplification of large signals. By 1922 they had reached the high point in their development and had almost entirely superseded crystal sets.

SOURCE: "The development of the art of radio receiving from the early 1920's to the present". by W.O.Swinyard. Proc.IRE (May 1962) p.794.

SEE ALSO: "Some recent developments of regenerative circuits" by E.H. Armstrong. Proc.IRE Vol.10. pp.244-260. August 1922.

ALSO: "The regenerative circuit" by E.H.Armstrong. Proc.Radio Club of America. April,1915.

1912   TUNGAR RECTIFIER   I.Langmuir (U.S.A.)

During the spring and summer of 1912, Dr.Irving Langmuir of General Electric became interested and active in the development of the gas-filled, tungsten-filament, incandescent lamp. These early laboratory lamps were constructed with heavy-coiled tungsten-spiral filaments that operated at low voltage. The bulb space around the filament was relatively small so that it would operate at high temperature and a few drops of mercury were placed in the bulb. There was usually another portion of the bulb where the mercury vapour condensed and ran back into the filament portion.

These lamps were frequently tested by operation from a 110 v. d.c. line through a series resistance. As one of the objects of these tests was to study the life and characteristics of the lamp at very high temperatures, there were frequent burnouts. It was noticed that, when the filament sometimes burned out during operations, an arc formed at the break.

In all of this early work the primary interest was in the arc as a source of light and apparently no tests on the arc as a rectifier were made at that time. However, one of the men sketched a tube with a hot filament and a separate anode plus liquid mercury.

In the early part of 1915 the possible need for a garage battery charger was considered and during 1916 it became a going concern.

SOURCE: "Early History of Industrial Electronics" W.C.White. Proc. IRE May 1962. p.1130.

1912   AERIALS (Ionospheric propagation)   W.H.Eccles (U.K.)

Following Marconi's success in 1901 in transmitting signals across the Atlantic, Kennelly and Heaviside postulated the existence of a conducting (ionized) layer in the earth's upper atmosphere and suggested that such a layer might cause the waves to follow the curvature of the earth. After it became clear that diffraction could not explain the substantial field strengths actually received at great distances, increased attention was directed to this proposal of an ionized region.

The theory of radio-wave propagation through the ionosphere is based on work by Eccles in 1912, on the ionizing effect of solar radiation, and on the effective refractive index of an ionized medium. Larmor in 1924 re-examined the work of Eccles and others and ascribed the major part of the refractive effect to the presence of free electrons in large numbers. The Eccles-Larmor theory, as later extended and developed by Appleton, Hartree, and others to include the effect of anisotropy due to the earth's magnetic field, is now considered the basic theory of radio-wave propagation in the ionosphere. This work was later extended by Booker and others to cover oblique propagation in a non-homogeneous ionosphere.

SOURCE: "Radio-wave propagation between World Wars I and II" S.S.Attwood Proc.IRE (May 1962) p.689

SEE ALSO: "On the diurnal variations of the electric waves occurring in nature and on the propagation of electric waves round the bend of the earth" W.H.Eccles Proc.Roy.Soc.(London) A. Vol:87. pp.79-99. June 1912.

1912   CIRCUITRY - Heterodyne and
        Superheterodyne   H.M.Fessenden
                                                    (U.S.A.)
                                   E.H. Armstrong

Professor Fessenden, in his search for an improved receiver, invented the heterodyne system in 1912. Previous receivers had merely acted as valves, detecting by turning a direct current on and off in amounts proportional to the received signal. In contrast, the heterodyne system operated through the joint action of the received signal and a local wave generated at the receiving station. Combination of these two alternating currents resulted in an audio beat-note, the difference frequency between the two waves. Although Fessenden's local oscillator was an arc source, very bulky and troublesome, it was nevertheless the forerunner of superheterodyne and single-banded reception.

The next advance in double-detection technique involved amplification of the beat-note or intermediate frequency. Several parallel developments took place in the United States and in Europe. It is difficult to name an inventor since the superheterodyne system as a basic idea seemed to appear from several sources at about the same time. The works of J.H.Hammon, A.Meissner, Lucian Levy, E.F.Alexanderson, and E.H.Armstrong stand out. Armstrong fully appreciated the problem and obtained a patent in 1920 that was of major importance in the practical application of the superheterodyne system.

SOURCE: "Radio Receivers - past and present" by C.Buff. Proc.IRE (May 1962) p.887.

SEE ALSO: "The superheterodyne - its origin, development and some recent improvements" by E.H.Armstrong. Proc.IRE Vol.12 (Oct.1924) p.549.

1912   CLOUD CHAMBER   C.T.R.Wilson   (U.K.)
(For revealing the ionisation tracks of radio-active particles)

His intention was to produce an artificial cloud by the adiabatic expansion of moist air. For condensation to occur it was then believed that each droplet required a nucleus of dust. However, Wilson showed that even in the complete absence of dust particles some condensation was possible, and that it was greatly facilitated by exposure to X-rays. He was led to conclude that charged atoms (or ions) were the necessary nuclei (1896-7). After much labour he produced, in 1911, his Cloud Chamber, in which the paths of single charged particles showed up as trails of minute water droplets.

The Wilson Cloud Chamber, sometimes modified or refined, has been an indispensable tool of modern physics ever since.

SOURCE: "A Biographical Dictionary of Scientists" T.I.Williams. Adam and Charles Black, London 1969. p.564.

1913   RELIABILITY (Standards)   AIEE (U.S.A.)

At the risk of some controversy, it can be stated that the first step in the new field of reliability can be traced to discussion generated by the AIEE Standard No.1, published in 1913. The title of this document was "General Principles upon which Temperature Limits are Based in the Rating of Electric Machines and Apparatus". This Standard started the first cycle of appreciation for the chemical composition of electrical materials. During the subsequent development of the electrical and electronic field, there has been a cyclic recurrence of this emphasis which can be recognized today as "analysis for the basic mechanism of failure in materials used in highly reliable electronic parts". However, in all fairness, it should be stated that the element of time-dependent degradation was not a part of the referenced first document. This element did not become a formally recognized factor until the great upsurge period for reliability in the 1950's.

SOURCE: "The reliability and quality control field from its inception to the present". by C.M.Ryerson. Proc.IRE (May 1962) p.1323.

1913   RESISTORS - (Thin metal film)   W.F.G.Swann (U.K.)

Since this is a review article it is necessary to refer to some of the earlier papers on the subject of thin films. A typical example is the paper that Dr.W.F.G.Swann presented to the British Association meeting over sixty years ago in 1913, which illustrates how little our ideas have changed in this period. Swann measured the resistivity of sputtered platinum films as a function of the sputtering time and hence of the film thickness; he also recorded the temperature coefficient of resistance of these films between liquid nitrogen temperature and 100°C and observed the abnormally high resistivity associated with very thin films and also a negative temperature coefficient of resistance in the thinnest films.

SOURCE: "Resistive thin films and thin-film resistors.-History, Science and and Technology" by J.A. Bennett. Electronic Components (Sept. 1964) p. 737.

SEE ALSO: W.F.G. Swann, Phil Mag. Vol. 28 (1914) p. 467.

1914 ULTRASONICS (Submarine detection) P. Langevin (France)

The introduction and development of ultrasonics spans a time of about 80 years. Ultrasonics (originally called supersonics, a name now used to describe speeds faster than sound) can be said to have started in 1883 with the Galton whistle. Capable of producing sound waves of a frequency higher than can be heard by humans, such whistles were used to demonstrate the reflection, refraction, interference and propagation of sound. A minor use was found in summoning dogs without alerting humans.

No further uses were found for ultrasonics until the advent of World War I, when an intensive program on submarine detection was carried out. Since sound waves are the only type of wave propagation that can detect submerged submarines, considerable effort was spent generating and detecting such waves. The most successful effort was carried out for the French government under the direction of P. Langevin. Using a sandwich made of quartz crystals cemented to steel plates, a directive sound beam was obtained which could be reflected off sunken objects and results were obtained which compared with those achieved up to the beginning of World War II. While this effort came too late to be of use in World War I, it laid the foundation for the extensive applications in World War II for both detecting and destroying submarines (by ultrasonic homing torpedoes) which contributed materially to Allied victory.

SOURCE: "Uses of Ultrasonics in Radio, Radar and Sonar Systems" by W.P. Mason. Proc. IRE (May 1962) p. 1374.

SEE ALSO: "Electroacoustics" by F.V. Hunt. Harvard Univ. Press. Cambridge, Mass. (1954) Chapter 1.

1914 THYRATRONS I. Langmuir (U.S.A.)

In 1914 Dr. Langmuir first suggested a method of controlling the arc in a mercy-pool tube by means of a grid. He showed how a grid voltage could be used to control the starting of the main arc in each rectifying circle. Thus, the average arc current through the tube was controlled when an a.c. anode voltage was used.

In 1922 Toulon, a French scientist, improved on this method of control by varying the phase of the grid voltage with respect to the anode voltage rather than to its amplitude. Thus, the arc could be made to start at any point in the anode voltage cycle and this resulted in a very practical and convenient method of controlling the average value of a rectified anode current.

In 1936 Dr. A.W. Hull developed the idea of operating a hot-cathode diode in a low pressure of an inert gas or vapour. As a result, the space charge effect was eliminated and the voltage drop of the discharge dropped to a low and more or less constant value of 5 to 10 volts. At this low arc drop voltage, the positive -ion bombardment of the cathode is not destructive.

SOURCE: "Early history of industrial electronics" W. C. White. Proc. IRE
May 1962. p. 1132.

1915   FILTERS (ELECTROMAGNETIC)   G. Campbell & K. W. Wagner   (U. K.)

        In 1831 Michael Faraday formulated the law of electromagnetic in-
duction and self-induction. Some 84 years later, in 1915, G. Campbell and
K. W. Wagner utilized Faraday's law in their invention of the first electro-
magnetic or LC wave filter. Significant advances in filter theory and technology
then followed rapidly, until today, filters have so permeated electronic technology that it is hard to conceive of a modern world without them. Consumer,
industrial, and military electronic systems all require some kind of signal
filter, and, in the past, LC filter networks provided one of the most efficient
and economical methods of implementing them.

SOURCE: "Inductorless filters: a survey" by G. S. Moschytz. IEEE Spectrum.
(August 1970) p. 30.
SEE ALSO: "The golden anniversay of electric wave filters" by A. I. Zverev.
IEEE Spectrum. Vol. 3. pp. 129-131 March, 1966.

ALSO: "Introduction to filters" by A. I. Zverev. Electro-Technol. (New York)
pp. 63-90. June 1964.

1915   RADIO (Single sideband communication)   J. R. Carson (U. S. A.)

        The military and international companies with a need to com-
municate abroad leaned heavily on radio, and by the 1950s designers were
hunting for ways to ease spectrum crowding. Single-sideband transmission
appeared to hold great promise.

        Like so many other radio techniques, it had been developed
originally for use in telephone systems. In 1915, J. R. Carson of AT & T's
Development and Research Dept. had proved that only one sideband was
needed for transmitting intelligence. Three years later the first commercial
application of single-sideband showed that it was possible to use this technique to increase channel capacity. By the mid-1930s single-side-band
transmission at high radio frequencies had proved successful. In the late
50s, after its first major application in the Strategic Air Command became
known, single-sideband really caught on.

SOURCE: "Communications" J. H. Gilder. "Electronic Design" 24.
November 23, 1972. p. 96.

1916   RELIABILITY (Control)   Bell/Western Electric   (U. S. A.)

        Perhaps one of the first clear-cut examples of what would now be
called a reliability-control program occurred in 1916. The Western Electric
Company and the Bell Telephone Laboratories co-operated in a planned program
to produce good performing and trouble-free telephone equipment for public use.
The following elements of a good reliability program were involved:-

(1)   An ably planned, forward looking R&D program considering the system needs.

(2)   Development programs leading to mature designs verified by design reviews and engineering model tests.

(3)   Part improvement projects using test-to-failure methods, consideration of tolerances and performance under stress loading.

(4)   Standardization and simplification in design.

(5)   Product evaluation under field application conditions for prototypes and pilot production models.

(6)   Quality control during manufacture to assure achieving the reliability inherent in the design.

(7)   Feedback from the field to provide information for the designers.

SOURCE: "The reliability and quality control field from its inception to the present" by C. M. Ryerson. Proc. IRE (May 1962) p. 1323.

1917   CONDENSER MICROPHONE   E. C. Wente   (U.S.A.)

One electrostatic device that has earned a secure place for itself in the transducer art is the condenser microphone. Wente's "uniformly sensitive instrument "of 1917 probably represents the first transducer design in which sensitivity was deliberately traded for uniformity of response, and it was almost certainly the first in which electronic amplification was relied on to gain back the ground lost by eschewing resonance.

SOURCE: "Electroacoustics" by F. V. Hunt. Harvard Univ. Press 1954. p. 169.

SEE ALSO: "The sensitivity and precision of the electrostatic transmitter for measuring sound intensities" by E. C. Wente, Phys. Res. Vol. 19. (May 1922) p. 498.

1917   CRYSTAL PULLING TECHNIQUE   J. Czochralski

Most of the single crystal silicon used for p-n junction devices is produced by one of three techniques. The oldest of these is the pulling process commonly called the Czochralski technique whereby the crystal is grown from a charge of Si melted in a crucible (usually quartz). Historically, the next development was the float-zone technique of crystal growth, involving the movement of a narrow molten zone up or down a vertical rod of initially polycrystalline silicon. This technique has been widely used because it eliminates the problem of crucible contamination. The third crystal growth technique, which has only relatively recently been commercially exploited, is a crucibleless pulling technique whereby a molten region is formed on the end of a thick silicon rod and the crystal is seeded and pulled from this molten pool in a manner combining, to some extent, features of Czochralski and float-zone growth.

SOURCE: "Microinhomogeneity problems in silicon" by H. F. John, J. W. Faust and R. Stickler. IEEE Trans. on Parts, Materials and Packaging. Vol. PMP-2 No. 3. September, 1966. p. 51.

SEE ALSO: "Measuring the velocity of crystallisation of metals" by
J. Czochralski. Z. Phys. Chem. Vol. 92 (April 1917) p. 219.

1918   CIRCUITRY (NEUTRODYNE)   L. A. Hazeltine (U. S. A.)

The neutrodyne circuit was invented in 1918 by Hazeltine. Basically this
was a tuned radio-frequency (TRF) amplifier which employed a specific type of
neutralization. A current obtained from the plate circuit was fed back into the
grid circuit in the proper magnitude and phase to balance out, or neutralize, the
effect of grid-to-plate capacitance inside the tube, and thus it achieved stability
and prevented oscillation.

SOURCE: "The development of the art of radio receiving from the early 1920's
to the present" by W. O. Swinyard. Proc. IRE (May 1962) p. 794.

SEE ALSO: "Tuned radio frequency amplification with neutralization of capacity
coupling" by L. A. Hazeltine. Proc. Radio Club of America. Vol. 2. March, 1923.

1918   AERIALS (Ground wave propagation)   G. N. Watson (U. K.)

At all frequencies, electromagnetic energy is diffracted into the
geometric shadow of an obstacle, the amount increasing with decreasing frequency. In the radio-frequency range, diffraction yields significant and usable
signal strengths in the shadow zone beyond the geometric horizon of the earth.
The mathematical problem of calculating the diffraction field strength in the
shadow zone of the earth was attacked by many mathematicians over 15 years
with resulting divergent answers. The first significant break--through was
produced by Watson in 1918, who showed that waves radiated by an antenna on
the surface of a perfectly conducting sphere would be attenuated exponentially
at great distances. The numerical values of field strength predicted by Watson
under these conditions proved to be far lower than known experimental values.
This discrepancy promoted increased interest in the ionosphere as a mechanism
that might explain the wide divergence in the mathematical and experimental
values of field strength, particularly in the kilocycle range of frequencies.

SOURCE:"Radio-wave propagation between World Wars I and II" S. S. Attwood
Proc. IRE (May 1962) p. 688.

SEE ALSO: "The diffraction of electric waves by the earth" G. N. Watson.
Proc. Roy. Soc. (London) A, Vol. 95. pp. 83-89. October 1918 and Vol. 96.
pp. 546-563. July 1919.

1918   MULTIVIBRATOR CIRCUIT   H. Abraham & E. Bloch   (France)

The multivibrator, described by Abraham and Bloch in 1918, is a very
important circuit, and is in use for many purposes. In its original form, however, it is more commonly used for the generation of pulses than as a time base
or discharger circuit.

The multivibrator, consists of two valves, each having the anode
coupled to the grid of the other via a condenser, and with a leak resistance to
earth. The method of operation of the circuit is as follows: when the anode
current in one valve increases, it passes a negative signal to the grid of the

other, due to the increased voltage drop across the anode load. The anode current of the second valve is thus reduced, which results in the grid of the first valve becoming more positive and increases the anode current of this valve still further. It will thus be apparent that the effect is a cumulative one and that the anode current in the first valve rapidly reaches a maximum while the anode current in the second is cut off. This circuit condition then remains until the charge in the condenser C_1 has leaked away through R_2 sufficiently to permit anode current to flow once more in the second valve. When this occurs, the cumulative effect again takes place, but in the reverse direction. The multivibrator, therefore, has two unstable limiting conditions which occur when either of the valves is at cut-off and the other at zero grid potential.

SOURCE: "Time Bases" by O. S. Puckle. Chapman & Hall London (1944) p. 25.

SEE ALSO: "Notice sur les Lampes-valves a 3 Electrodes et leurs applic-ations" Publication No:27 of the French Ministere de la Geurre (April 1918)

1918   NOISE (Shot effect)   W. Schottky   (Germany)

       The first realisation that unwanted random noise was a factor to contend with in the field of communications came during World War I when attempts were being made to design high-gain vacuum-tube amplifiers. It was soon found that there was a limit to the number of stages which could be cas-caded in the quest for high gain due to an unacceptably high background noise which masked the weak signals being amplified. In his classic paper Schottky first explained one of these effects and formulated the random component in the plate current of a vacuum tube.

       Schottky ascribed the random fluctuations in the plate current to the fact that this current is composed not of a contunuum but rather of a sequence of discrete increments of charge carried by each electron arriving at the plate at random times. The average rate of charge arrival constitutes the dc compo-nent of the plate current on which is superimposed a fluctuation component as each discrete charge arrives. He referred to this phenomenon as "schroteffekt" or "shot-effect".

SOURCE: "Noise and Random Processes" by J. R. Ragazzini and S. S. L. Chang Proc. IRE (May 1962) p. 1146.

SEE ALSO: "Theory of shot effect" by W. Schottky. Ann. Phys. Vol. 57 (Dec. 1918) p. 541.

1918   INDUCTION HEATING (High Frequency)   E. F. Northrup   (U. S. A.)

       The production of heat by induced currents was recognised as early as 1880. The heating of transformer cores due to eddy currents was, of course, understood at an early date. Probably the real engineering and developmental pioneer in induction heating with frequencies above the power range was Prof-essor E. F. Northrup of Princeton University. In 1918 he built practical "furnaces" for frequencies above 10,000 cycles and powers as high as 60 kw using a spark gas oscillator.

SOURCE: "Early history of industrial electronics" W. C. White. Proc. IRE May 1962. p. 1134.

1919   RESISTOR (Metal-film type)   F. Kruger (Germany)

        Proposals for the production of metallic-film resistors that will possess a high degree of stability date back over 25 years, but mostly they could be prepared only with comparatively low resistance values. By spiralling the coating in a manner similar to that used for high-stability resistors, it has been found possible to increase the resistance of resistors made with gold films up to 200 000 ohms.

SOURCE: "Fixed resistors for use in communication equipment" by P. R. Coursey. Proc. IEE Vol. 96 Pt. III (1949) p. 173.

SEE ALSO: F. Kruger British Patent No: 157909/1919.

1919   CIRCUITRY (Flip-Flop Circuits)   Eccles & Jordan   (U.S.A.)

        In digital circuitry probably the most important single device is the bistable circuit or "flip'flop". Used today in innumerable modifications the original form of this circuit was developed in 1919 by Eccles and Jordan. During the late 1930's flipflops in tandem connection began to be used by nuclear physicists to scale down the counting rate from Geiger-Muller counters by factors of 2^n. These devices were all essentially binary counters in which the last "carry" pulse was used as the output to a mechanical register. In 1944, Potter described a scale-of-ten counter in which feedback caused a binary counter to skip six of its possible 16 states. A neon lamp indicating the state of each flip-flop provided a binary-coded readout for each decade. In 1946, Grosdoff described another binary counter with feedback to provide a scale-of-ten circuit. Grosdoff's circuit featured a direct readout obtained by a matrix of resistors arranged to light one of ten neon lamps for each stablestate of the counter. By connecting such decades in tandem direct reading decimal counting registers of arbitrarily large capacity were achieved and the way was cleared for the widespread application of high speed electronic counting.

SOURCE: "Digital display of measurements in instrumentation" B. M. Oliver Proc. IRE May 1962   p. 1170.

1919   RETARDED FIELD MICRO-   H. Barkhausen & K. Kurtz (Germany)
         WAVE OSCILLATOR.

        In 1919 it was noted by Barkhausen and Kurtz, during tests for the presence of gas in transmitting valves in which the grid was held at a high positive potential and the anode at a negative one, that oscillation could be maintained in a circuit connected between grid and anode, or between other pairs of electrodes.

        The explanation was given that electrons are accelerated from the cathode to the positive grid through which some of them pass. These are then retarded in the grid anode space and turn back to the grid, some of them again passing through, and being reflected at the cathode to repeat the behaviour.

SOURCE: "Microwave Valves; A survey of evolution, principles of operation and basic characteristics" by C. H. Dix and W. E. Willshaw. Journal Brit. IRE August, 1960. p. 578.

SEE ALSO: H. Barkhausen and K. Kurtz "The shortest waves obtainable with vacuum tubes" Z. Phys. Vol. 21 p. 1-6, 1920.

1919   VALVES - HOUSKEEPER SEAL   W. G. Houskeeper (U.S.A.)

The breakthrough came in 1919 when W. G. Houskeeper patented a method of sealing base metals through glass. Before this the leads brought through the glass had to be wires or ribbons. With this invention it was possible to make large diameter seals using metals such as high purity copper which has high conductivity, and is easily worked into any desired shape. The expansion of copper is $165 \times 10^{-7}/°C$ and that of a common glass $52 \times 10^{-7}/°C$ so a matched seal is not possible. The technique developed by Houskeeper was to taper the metal to a feather edge and then seal the glass to the thin edge of the taper. The copper is cleaned, oxidised and then sealed to the glass. The thin copper is sufficiently flexible to equalise the difference in expansion and contraction between the glass and metal.

With this development it became possible to make the anode part of the vacuum envelope and cool its outer surface directly. By 1925 valves with water cooled anodes were in commercial production and they mark the beginning of the era of high-power broadcast transmitters.

SOURCE: "Electronics Engineer's Reference Book" Newnes-Butterworth London (1976) Chap. 7. p. 7 - 31

SEE ALSO: "The art of sealing base metals through glass" by W. G. Houskeeper. Jour. Amer. Inst. EE Vol. 42 (1923) p. 870.

1919   TELEVISION (Electronic System)   Vladimir Zworykin (U.S.A.)
1927   TELEVISION (Electronic System)   Philo Farnsworth (U.S.A.)

It was already recognised that a complete cathode-ray system provided the answer, but no experimental work was started and the idea lay dormant until, after going to America in 1919, Zworykin joined the Westinghouse Company. His major difficulty centred on the electronic camera tube. He, like Campbell-Swinton, had conceived of the idea of charge storage by 1919. Zworykin lacked the necessary funds to carry his ideas forward into practical form at the time he conceived them, and it was several years before he could concentrate on their elaboration. Westinghouse took him on to their research staff but their laboratory devoted itself mainly to radio research, and, since Zworykin was given no freedom to pursue his ideas on television, he resigned to join a development company in Kansas. Returning to Westinghouse in 1923, he drew up an agreement whereby he retained the rights to the television inventions he had disclosed in 1919, while Westinghouse acquired the exclusive option to purchase the patents at a later date.

Philo Farnsworth was essentially an individual inventor who, though fortunate enough to find substantial financial backing, always retained his autonomy in research. Largely self taught, he appears at an early age to have conceived of a completely electronic system. He was of the type which prefers to work on a small scale with relatively simple equipment. Working in laboratories in Los Angeles and later in San Francisco he was able to

demonstrate a complete electronic system in 1927, when he filed his first
patent application, including his image dissector tube, which constitutes his
most important inventive contribution. After long-drawn-out patent interference proceedings, Farnsworth and Zworykin each received basic patents
on their different systems of television transmission.

SOURCE: "The Sources of Invention" by J. Jewkes, D. Sawers and
R. Stillerman. MacMillan & Co. London 1958. p. 385/6

SEE ALSO: "The History of TV" by G. R. M. Garratt and A. H. Mumford.
Proc. of Institution of Electrical Engineers, Part III A, Television, vol. 99,
1952.

1920   ULTRA-MICROMETER   R. Whiddington (U.K.)

        Few would suspect that one of the first authentic disclosures of pract-
ical electronic instrumentation was published in the November 1920 issue of
The Philosophical Magazine by R. Whiddington M.A., D.Sc., Cavendish Professor of Physics in the University of Leeds.
        The title of Professor Whiddington's communication is "The Ultra-
Micrometer; an Application of the Thermionic Valve to the Measurement of
Very Small Distances". The summary of the paper reads: "If a circuit consisting of a parallel-plate condenser and inductance be maintained in oscillation by
means of a thermionic valve, a small change in distance apart of the plates
produces a change in the frequency of the oscillations which can be accurately
determined by methods described. It is shown that changes so small as 1/200
millionth of an inch can easily be detected. The name "ultra-micrometer" is
tentatively suggested for the apparatus".

SOURCE: Letter: "A pioneer of electronics" by F. G. Diver. The Radio and
Electronic Engineer. Vol. 45. No 11 (Nov. 1975) p. 687.

SEE ALSO: "The ultra-micrometer - an application of the thermionic valve to
the measurement of very small distances" by R. Whiddington. Phil. Mag. Series 6
Vol. 40 (Nov. 1920) p. 634.

1921   SHORT WAVE RADIO   Amateurs   (U.S.A.) and (Europe)

        One of the main developments in radio after 1918 was the discovery of
the usefulness of the shorter wave-bands. It was generally considered that wave-
lengths below 200 metres were useless except for short distance transmission,
though cases were known of long ranges being obtained on short waves. These
were regarded as freaks, however, and wave-lengths below 200 metres were,
after 1918, allocated to amateurs who encouraged by these 'freak' results,
arranged trial broadcasts from America to England in December 1921 on 200
metres. Their success showed that short-wave low-power broadcasts could be
heard over long distances.

SOURCE: "Sources of Invention" by J. Jewkes, D. Sawers and R. Stellerman.
MacMillan. London 1958. p. 353.

1921   <u>FERROELECTRICITY</u>   J. Valasek

Ferroelectricity was first discovered by Valasek in 1921 in Rochelle salt, which is piezoelectric above its Curie point; however, the most important ferroelectric material today is barium titanate, which was discovered in 1942 and developed shortly thereafter at M.I.T. for ferroelectric applications. Since then, other materials of the Perovskite structure and of other structures have been found to be ferroelectric.

<u>SOURCE</u>: "Solid State devices other than Semiconductors" by B. Lax and J.G. Mavroides. Proc. IRE (May 1962) p. 1014.

<u>SEE ALSO</u>: "Piezoelectric and allied phenomena in Rochelle salt" by J. Valasek. Phys. Rev. Vol. 17. (April 1921) p. 475.

<u>ALSO</u>: "High dielectric constant ceramics" by A. von Hippel, R. E. Breckenridge, F. E. Chesley and L. Tisya. Ind. Engrg. Chem. Vol. 38 (Nov. 1946) p. 1097.

1921   <u>CRYSTAL CONTROL OF FREQUENCY</u>   W. G. Cady   (U.S.A.)

While working with a quartz crystal connected in the circuit of a self-excited vacuum-tube oscillator, Cady discovered that the frequency of self-oscillation could be stabilized over a small range by the vibration of the quartz crystal. In retrospect it appeared that a good many other experiments had made use of crystals associated with vacuum-tube circuits with which they might have observed this stabilizing action; but most of them sought to avoid the "anomalous" effects that occurred in the neighbourhood of resonance in the crystal, and it remained for Cady to discover this remarkable stabilizing action of a high-Q resonator. Cady extended these results by applying two pairs of terminals to the crystal and connecting these as a feedback path for a three-tube amplifier in such a way that the circuit would oscillate but only at the resonance frequency of the crystal.

Cady described and demonstrated his crystal circuits (January 1923) to Professor G. W. Pierce of Harvard University. Pierce took up the study of such crystal-oscillator circuits immediately, and within a few months his experiments had led him to the invention of several improved forms of crystal oscillator in which a two-terminal crystal could be made to control uniquely the frequency of oscillation in a single-tube circuit.

<u>SOURCE</u>:   "Electroacoustics" by F. V. Hunt. Harvard Univ. Press (1954) p. 53 & 55.

<u>SEE ALSO</u>:   "The Piezo-electric resonator" by W. G. Cady. Phys. Rev. 17, 531 (A) (April 1921). Proc. Inst. Radio Engrs. 10, 83-114 (April 1922); also Piezoelectricity (New York and London, McGraw-Hill Book Co., Inc., 1946).

<u>ALSO</u>:   Walter G. Cady, (crystal resonator) U.S. Patent No: 1,450,246 (filed 28 January 1920) issued 3 April 1923; also (crystal stabilization, and a 3-tube oscillator controlled by a 4-terminal crystal) U.S. Patent No: 1,472,583 (filed 28 May 1921) issued 30 October 1923.

1922   NEGATIVE RESISTANCE   E.W.B.Gill and J.H.Morrell   (U.K.)
OSCILLATOR

Following the work of Barkhausen and Kurtz (1919) which for the
first time reported the generation of electrical oscillations depending primar-
ily on the oscillatory motion of electrons in a vacuum, and not on the excitation of oscillatory currents in a tuned circuit, and this discovery might be
said to represent the starting point of the whole field of modern microwave
valves.

Subsequent work on this device called generally the "retarding
field" generator, by Gill and Morrell, reported in 1922 and subsequently by
many others, showed that this simple type of operation had many variants
including one in which with adequate emission, a negative resistance could be
provided by the tube over a frequency band.

SOURCE: "Microwave Valves:A survey of evolution, principles of operation
and basic characteristics" by C.H.Dix and W.E.Willshaw. Journal Brit.IRE
August,1960. p.578.

SEE ALSO: "Short electric waves obtained by valves" by E.W.B.Gill and
J.H.Morrell . Phil.Mag. 44, p.61. 1922.

1923   ICONOSCOPE   V.K.Zworykin (U.S.A.)

The combination of electron beam scanning and storage was first proposed and carried into practice by V.K.Zworykin in 1923 - at a time preceding
anh other electronic pickup system. The target was an aluminium film oxidised
on one side, which was photosensitized with cesium vapor and faced a metal
grill serving as collector for photoelectrons.   The metal side, which served as
signal plate, was scanned by a high-speed electron beam, which penetrated
through the oxide layer, forming a temporary conducting path permitting the
locally stored charge to flow off through the signal plate.

While the tube as described was capable of transmitting only rudimentary patterns, it became the ancestor of the extended line of storage camera
tubes which dominate all phases of television picture transmission. Their
common features are electron beam scanning and a picture target with small
transverse conductivity capable of storing charge released in response to light.

The first practical storage camera tube was the iconoscope. Here, the
picture was projected on a mosaic of photosensitive elements capacitatively
coupled to a signal plate. The mosaic was then scanned by a high-velocity beam,
restoring the mosaic to a uniform potential and releasing photoelectrically
stored charge for forming the picture signal. The secondary-emission ratio was
greater than unity, so that the equilibrium potential under the beam was close to
that of the electrodes facing the mosaic.

SOURCE: "Beam-deflection and photo devices" by K.Schlesinger and E.G.
Ramberg. Proc.IRE (May 1962) p.993.

SEE ALSO: V.K.Zworykin   U.S.Patent No 2,141,059 (Dec.20th,1938) filed
December 29th,1923.

ALSO: "The Iconoscope - a modern version of the electric eye" by V.K.
Zworykin. Proc.IRE Vol.22. (Jan.1934) p.16.

1924   <u>LINEAR SAW-TOOTH TIME</u>   R. Anson   (U.K.)
            <u>BASE CIRCUIT</u>

          The earliest attempt at the construction of a linear saw-tooth time base
is believed to be due to Anson, and was developed about the year 1924. The
author believes that this instrument appeared as a result of the development
of the Anson relay in which a neon tube was used for signal-shaping purposes in
conjunction with telegraph receivers. The neon tube is a two-electrode valve,
filled with neon gas at a low presure, in which neither of the electrodes is heated.
When a potential, which is dependent upon the gas pressure, the proximity of
the electrodes, the material of which the electrodes are made and their surface
condition, is applied between the electrodes the gas becomes ionized. This
potential is normally about 130 volts. When the potential falls to about 100 volts
the gas deionizes.

          The neon time base consists of a condenser charged through a high
resistance and discharged by the neon tube when the charge upon the condenser
reaches the striking voltage of the tube.

<u>SOURCE</u>:   "Time Bases" by O.S. Puckle. Chapman & Hall, London (1944) p. 13.

<u>SEE ALSO</u>:   R. St. G. Anson, British Patent No: 214754.

1924/5   <u>RADAR</u>   E. Appleton, G. Briet, R. A. Watson Watt
                           S.F.R.,   G.E.M.A. et al   (primarily U.K.)

          The first use found for the reflecting properties of radio waves was in
measuring the height of the Heaviside layer. This was done in Britain in 1924
by Sir Edward Appleton and M.F. Barnett, and in the U.S.A. in 1925 by Dr.
Gregory Breit and Dr. Merle A. Tuve of the Carnegie Institute. Breit and Tuve
were the first to apply the pulse principle.

          By 1939 Germany, Great Britain, Holland and the U.S.A. all posses-
sed military radar apparatus, while the first peaceful application had been made
in France in 1935. In France and Germany the work was done by the scientists
of radio companies: in Britain and the U.S.A. by scientists in government
research stations.

          Scientists of the Societe Francaise Radioelectrique began to study the
use of metric and decimetric radio waves to detect obstacles in 1934, first with
a view to saving life at sea and then for military uses. An 'obstacle detector'
working on decimetric waves and employing magnetrons and the pulse principle
was fitted to the liner Normandie in 1935; it appears to have been successful
and equipment was installed to detect ships entering and leaving the harbour at
Le Havre in 1936. Work was meanwhile done on equipment for detecting aircraft, though no radar warning system had been installed by 1939.

          German work, begun before 1935, was carried on, under contracts
from the Navy, by a new form, G.E.M.A., and later by other firms. As a
result, by 1939 a large number of radar sets were in operation for detecting
aircraft. Though the German work began at least as early as that in other
countries and had reached a very similar level of development by 1939, it
lagged behind after the outbreak of war. German policy was based on the

assumption that the war would be short, and consequently less effort was put into such basic work as radar than in Britain or the U.S.A.

Interest in the possibility of using radio waves to detect aircraft arose in Britain about 1934. The reflection of radio waves from aircraft had been observed in 1931 and 1933; on the latter occasion the possibilities for aircraft detection were carefully analysed. H.E. Wimperis, Director of Scientific Research at the Air Ministry and his assistant, A.P. Rowe, suggested that the country should increase its efforts to develop a method of detecting aircraft at a distance and an investigating committee of three scientists, Sir Henry Tizard, Professor A.V. Hill and Professor Patrick Blackett, was set up. It was soon realised that radio beam would be the ideal alternative to the existing inadequate acoustic warning equipment.

Robert A. Watson-Watt (now Sir Robert) who was a lecturer in physics at University College, Dundee, before he began a research career in the government laboratories, played the major role in developing practical radar equipment in Britain. He was superintendent of the Radio Division of the National Physical Laboratories at the time the pressure for improved air defence reached its peak. He felt confident that radio waves could be employed to detect aircraft. His two memoranda of February 1935 described his suggested means for so using them; after a demonstration of the echoes produced by aircraft from the B.B.C. Daventry short-wave station, the Tizard Committee recommended that work on the lines suggested by Watson-Watt should be started. Working with six assistants, of whom A.F. Wilkins was the principal, Watson-Watt developed the first practical radar equipment for the detection of aircraft on the Suffolk coast in the summer of 1935. The main problems he solved were the construction of a high-power transmitter, the modulation of it with short pulses, the development of receivers to handle the pulses and of suitable transmitting and receiving aerials. The performance of the first equipment was considered promising enough for the Air Ministry to build a chain of five radar stations.

The development of radar had meanwhile been proceeding independently in the United States. Military interest began after L.A. Hyland, an associate of A.H. Taylor, discovered accidentally in 1930 that aircraft cause interference in radio waves and Leo Young successfully applied the pulse apparatus to this. Despite the fact that radar looked so promising, the Navy was reluctant to spend any significant amount on it, but, through the persistent efforts of Harold G. Bowen, chief of the Naval Laboratory, $100,000 was allocated for radar research. Robert M. Page, head of the research section of the Naval Laboratory's radio division, developed some of the first modern radar equipment. In 1938, two years after successful laboratory demonstrations of the equipment, the American Navy finally fitted radar devices to some of its ships.

After 1940 Great Britain and the United States co-operated in radar development.

SOURCE: "The Sources of Invention" by J. Jewkes, D. Sawers and R. Stillerman. MacMillan & Co. London 1958. p. 346, 347 & 348.

1925   <u>ELECTROSTATIC LOUDSPEAKER</u>   -   (Various)

   The electrostatic loudspeaker failed to gain wide commercial acceptance, in spite of extensive development activity devoted to it during 1925-1935,
for the very sound reason that several serious shortcomings still adhered to its
design.   Either the diaphragm or the air gap itself had usually been relied on to
provide the protective insulation against electrical breakdown, but this protection was often inadequate and limits were thereby imposed on the voltages that
could be used and on the specific power output.   Close spacings, a film of trap-
ped air, stiff diaphragm materials, and vulnerability to harmonic distortion
combined to restrict to very small amplitudes both the allowable and the attain-
able diaphragm motion; and as a consquence, large active areas had to be
employed in order to radiate useful amounts of sound power, especially at low
frequencies.   But when large areas were employed, the sound radiation was
much too highly directional at high frequencies.   Several of the patents cited
below bear on one or another of these features, and it is now apparent that an
integration of such improvements would have made it possible to overcome
almost - but not quite - every one of these performance handicaps.   Occurring
singly as they did, however, no one of these good ideas was able by itself to
rescue the electrostatic units from the burden of their other shortcomings.
Taken together, however, with the newly available diaphragm materials and
with the important addition of one or two new ideas, the modern form of electrostatic loud-speaker can so completely surmount these former handicaps that
it merits careful consideration as a potential competitor for the moving-coil
loudspeaker in many applications.

<u>SOURCE</u>: "Electroacoustics" by F. V. Hunt. Harvard Univ. Press (1954)
p. 173 & 174.

<u>SEE ALSO</u>: For example, Colin Kyle, U. S. Pats. No:1, 644, 387 (filed 4 October
1926) issued 4 October 1927, and No:1, 746, 540 (filed 25 May 1927) issued 11
February 1930; Ernst Klar (Berlin) German Pat. No:611, 783 (filed 22 May 1926)
issued 5 April 1935, and U. S. Pat. No:1, 813, 555 (filed 21 May 1927, renewed 14
November 1930) issued 7 July 1931 (insulating spacers, perforated plate coated
with a dielectric); Hans Vogt (Berlin), more than a score of contemporary and
relevant German patents, for example, German Pats. No:583, 769 (filed 25
December 1926) issued 9 September 1933 and No:601, 117 (filed 17 May 1928)
issued 8 August 1934, and U. S. Pat. No:1, 881, 107 (filed 15 September 1928)
issued 4 October 1932 (tightly stretched diaphragm between perforated rigid
electrodes): Edward W. Kellogg (G. E. Co.) U. S. Pat. No:1, 983, 377 (filed 27
September 1929) issued 4 December 1934 (sectionalized diaphragm with induct-
ances for inpedance correction): William Colvin, Jr., U. S. Pat. No:2000, 437
(filed 19 February 1931) issued 7 May 1935 (woven-wire electrodes): D. E. L.
Shorter, British Pat. No:537, 931 (filed 21 February 1940, complete spec. 23
January 1941, accepted 14 July 1941) (diaphragm segmentation with external
dividing networks for improving directivity and impedance).

<u>ALSO</u>: "Wide range electrostatic loudspeakers" by P. J. Walker. "Wireless
World" Pt. 1. (May 1955 p. 208, Pt. 2 (June 1955) p. 265, Pt. 3 (August 1955)p. 381.

1925   RESISTOR (Cracked-carbon type)   Siemens and Halske   (Germany)

It was undoubtedly in Germany that the first practical use was made of a cracked-carbon film in lieu of metal to form a highly stable resistance coating and resistors of this general type were manufactured by several firms in that country for a number of years before the war. Amongst these, the Siemens and Halske organisation seems to have produced the largest quantities, so that this type of resistor commonly came to be known as the "Siemens resistor".

One of the earliest disclosures of the cracking of hydrocarbon vapour to produce a hard carbon layer is contained in the Seibt patent of 1930 and in the Stemag patent of the same year. The Siemens and Halske patent relating to these resistors is dated March 1932, but the fundamental method is already there referred to as "well known".

SOURCE: "Fixed resistors for use in communication equipment" by P.R. Coursey. Proc. IEE Vol. 96. Pt. III (1949) pp. 174-175.

SEE ALSO: Siemens and Halske Akt. Ges. German Patent No:438429/1925.

ALSO: Siemens and Halske Akt. Ges. British Patent No: 387150 (1932).

ALSO: C.A. Hartman. German Patent No: 438,429 (1925)

1925   NOISE (Johnson)   J.B. Johnson   (U.K.)

In addition to fluctuation effects produced by vacuum tubes, it was found that random noise signals were generated in metallic resistors made of homogeneous materials. These effects were found to be temperature dependent and are known as thermal noise. A number of basic contributions to the understanding of thermal noise were made in the 1920's and 1930's among which was the outstanding paper by Johnson in 1925. The source of thermal conductor noise was traced to the random excitation of the electron gas in the conductor in consquence of its existence in an environment of thermally-agitated molecules. The effect is similar to the Brownian movement of particles suspended in a liquid in which the thermally-agitated molecules of the liquid collide with the suspended particle and impart to it a certain amount of energy. Since the particle is cohesive, collision with any one of its molecules sets the entire particle in motion thereby resulting in random movements observable under a microscope.

SOURCE: "Noise and Random Processes" by J.R. Ragazzini and S.S.L. Chang Proc. IRE (May 1962) p.1147/8.

SEE ALSO: "Thermal agitation of electricity in conductors" by J.B. Johnson Phys. Rev. Vol. 32. 2nd Series. (July 1928) p. 97.

1925   SHORT WAVE COMMERCIAL RADIO
        COMMUNICATION   L.J.W. van Boetzelaer   (Holland)

On April 23rd, 1925, an experiment began in Hilversum which turned out to have far-reaching consequences for the link between The Netherlands and the Dutch East Indies, now Indonesia, causing a huge long-wave transmitter to become obsolete only months after it came into operation.

At the Nederlandsche Seintoestellen Fabriek, now PTI, a young research engineer, L. J. W. van Boetzelaer had just received a new water-cooled 4 kW transmitting triode from the Philips factory at Eindhoven, which had a grid-anode capacity low enough to permit oscillating at 11.5 MHz(!) Working in a humble wooden shed, Mr. van Boetzelaer, after many difficulties, managed to get the primitive transmitter "on the air".

To see how far this transmitter could be heard, it was arranged that the steam ship 'Prins der Nederlanden', sailing to the East, would listen in on a daily schedule. With admirable perseverance, van Boetzelaer operated the morse key until late at night, hoping to be read. The ship's reactions, coming in slowly via coastal stations, were favourable. Then, in a bold mood, the diligent operator invited Malabar to send a cable if they happened to read the transmissions. The telegram sent back caused great excitement in Hilversum. It flatly stated that reception on 26 metres had been loud and clear since the beginning of the experiment.

SOURCE: Philips Telecommunication Review. Vol. 33. No. 4. December 1975. page 191.

1925   TELEVISION (Mechanical scanning)   J. L. Baird (U.K.)

When Baird, in 1923, decided to devote his untried inventive genius to the development of a practical television scheme, the problem seemed to him to be comparatively simple. Two optical exploring devices rotating in synchronism, a light-sensitive cell and a controlled varying light source capable of rapid variations in light flux were all that were required, 'and these appeared to be already, to use a Patent-Office term, known to the art'. Baird, however, realised the difficult nature of the problem. "The only ominous cloud on the horizon", he wrote "was that in spite of the apparent simplicity of the task no-one had produced television".

Baird's principal contemporaries in this challenge were C. F. Jenkins of the U.S.A. and D. von Mihaly of Hungary. Other inventors were patenting their ideas on television at this time (1923) but only Jenkins, Mihaly and Baird and a few others were pursuing a practical study of the problem based on the utilization of mechanical scanners.

SOURCE: "The first demonstration of television" by Prof. R. W. Burns. Electronics & Power. 9th October, 1975. p. 953.

1926   FILM SOUND RECORDING
(Sound-on-disc system)   Warner Bros. (U.S.A.)

As a result of lagging interest in the motion pictures by the public, Warner Bros. in 1926, decided to test the popularity of sound pictures. To minimize the cost of the venture, this studio arranged to have Western Electric develop the necessary equipment to synchronize disk-recording machines with cameras that were housed in booths to suppress the camera noise. Arrangements were made with the Victor Talking Machine Company to do the recording in their facilities and with their personnel. The Victor Talking Machine Company was a Western Electric licensee, and their studios were equipped with

Western Electric recording equipment.

Western Electric developed motor drives for theatre projectors and disc turntables. These were mechanically connected to the same constant speed motor system. Essentially standard public address system amplifiers and loudspeakers were used. The first picture produced was "Don Juan". In October, 1927 "Jazz Singer" followed and was a success.

The public reaction was so enthusiastic that the large theatre chains wanted equipment immediately to play the pictures. Western Electric agreed to lease equipment to them. As a result of the success of the first pictures, Warner Bros. installed disk-recording equipment in their studios. This system was called Vitaphone. It was destined to be supplanted by systems that recorded the sound as photographic images on the same film the picture was printed on. Having demonstrated the popularity of sound pictures and developed the equipment, the industry proceeded with great speed to convert studios for sound-picture production.

SOURCE: "Film Recording and Reproduction" by M. C. Batsel and G. L. Dimmick. Proc. IRE (May 1962) p. 745.

1926   CIRCUITRY (Automatic volume control)   H. A. Wheeler (U. S. A.)

In 1925 the stage was set for the invention of a practical automatic volume-control circuit, and on January 2, 1926, Wheeler invented his diode AVC and linear diode-detector circuit. This circuit was first incorporated in the Philco Model 95 receiver which he designed at the Hazeltine laboratory and which was announced about September, 1929.

Full AVC bias voltage was applied to the first two RF tubes, and, to prevent distortion in the third RF stage, half AVC bias voltage was applied to that stage. The automatic volume-control action was sufficiently gradual to permit accurate tuning by ear, and it was unnecessary to touch the volume control once it was adjusted.

SOURCE: "The development of the art of radio receiving from the early 1920's to the present" by W. O. Swinyard. Proc. IRE (May 1962) p. 795.

SEE ALSO: "Automatic volume control for radio receiving sets" by H. A. Wheeler. Proc. IRE Vol. 16. (Jan. 1928) p. 30.

1926   YAGI AERIAL   H. Yagi (Japan)

Dr. H. Yagi studied under the direction of Prof. Dr. G. H. Barkhausen at the Dresden Technische Hochschule from 1913 to 1914, under the direction of Prof. Dr. J. A. Fleming at University College, London from 1914 to 1915, and under the direction of Prof. Dr. G. W. Pierce at Harvard University, Cambridge, Massachusetts from 1915 to 1916.

In 1926, during his career as a professor at Tohoku University, he invented the VHF directive "Yagi antenna" which was widely put into practical use for domestic television reception. Some years later, as a result of this invention, the Academy of Technical Science in Copenhagen awarded him the Valdemar Poulsen Gold Medal "for his outstanding contributions to radio technique."

SOURCE: "Death of Dr. Hidetsugu Yagi" Telecommunication Journal. Vol. 43. V/1976 p. 372.

1926   FIXED RESISTOR (Sprayed Metal Film)   S. Loewe (Germany)

         The basic idea involved is the very old art of decorating chinaware with precious metals. In Germany in 1926, Loewe developed a resistive film by atomizing a liquid solution of platinum resinate by forcing compressed air through it and applying the spray to an insulating base. Heating the film thus formed reduces it to the metal.

SOURCE: "Resistors - a survey of the evolution of the field" J. Marsten. Proc. IRE (May 1962) p. 922.

SEE ALSO: S. Loewe, German Patent No: 591, 735 (1926) and U. S. Patent 1, 717, 712 (1926)

1926   SCREENED GRID TUBE   H. J. Round (U. K.)

         By the end of the 1914-18 war the triode was the only tube in common use as a detector and amplifier and generator of high-frequency oscillations. Broadcasting was, in fact, started with the triode; although some of its limitations were, by that time, well recognised. One major limitation arose from the inherent electrostatic capacitance between the grid and the anode, within the valve itself. This gave rise to a coupling between grid and anode circuits which resulted in uncontrollable, and therefore undesirable, reaction between the output circuit and the input circuit. The introduction of a screen grid, between the control grid and the anode, to reduce this inter-electrode coupling, was first suggested by A. W. Hull. Schottky had earlier suggested a four-electrode tube, but his suggestion of the introduction of an additional grid had been to secure an increase of amplification factor. Hull's suggestion, by contrast, was directed to the reduction of grid-anode capacitance. But it remained for H. J. Round to bring the screen-grid valve into practical use in 1926.

         Other, and later, versions of the screen-grid valve were developed, the screen grid being provided with one or more skirts which extended to the walls of the container bulb. In one form the grid, and in another the anode, was brought out uniquely at the top of the tube, the other electrodes being brought out from the base. In this way the undesired capacitance between the control grid and the anode could be reduced to 0.001 or 0.01$\mu\mu$F.

SOURCE: "Thermionic devices from the development of the triode up to 1939" by Sir Edward Appleton. IEE Pub. Thermionic Valves 1904-1954. IEE London 1955. p. 22-23.

1927   NEGATIVE FEEDBACK AMPLIFIER   H. S. Black (U. S. A.)

         In one of the most fundamental discoveries in the history of communications, H. S. Black in 1927 at Bell Laboratories found that by feeding part of an amplifier's output back into its input (negative feedback), it was possible by sacrificing some amplification to achieve stable operation at low distortion.

SOURCE: "Mission Communications - the story of Bell Laboratories" by Prescott C. Mabon. Published by Bell Laboratories Inc., Murray Hill, New Jersey U. S. A. (1975) p. 171-2.

1927   FILM SOUND RECORDING   Fox Movietone News (U.S.A.)
        (Sound-on-film system)

        The first commercially successful photographic sound recording sys-
tem (Fox Movietone News) used a variable intensity method of modulating a
beam of light to expose the film negative. The gas-filled lamp known as the
Aeo-light had an oxide coated cathode, and its intensity could be modulated
over a considerable range by varying the anode voltage, at audio frequencies,
between 200 and 400 volts. The Aeo-light was mounted in a tube which entered
the camera at the back. Directly against the film was a light restricting slit
which passed a beam about a tenth inch long and 0.001 inch high, placed between
the picture and the sprocket holes. The Aeo-light could produce a sufficiently
high intensity to expose the sensitive negative films used for picture taking.
The system worked quite well for news photography, where the sound and picture were taken simultaneously on the same camera.

SOURCE: "Film Recording and Reproduction" by M. Batsel and G. L.
Dimmick. Proc. IRE (May 1962) p. 746.

1928   AERIALS (Diversity reception)   H. A. Beverage
                                        H. O. Peterson   (U.S.A.)
                                        J. B. Moore

        Because of the turbulence and abrupt changes encountered in the HF
medium, special attention had to be given to improved means of reception as it
became apparent that transmitter power increases were not alone sufficient.
        Among the significant techniques developed, one of the most important
is diversity reception, wherein a considerable improvement is obtained due to
the statistical independence in the fading characteristics of two or more paths.
Diversity reception is basic, improving the reception of any type of modulation
at any frequency. H. H. Beverage, H. O. Peterson and J. B. Moore described and
developed a triple-space-diversity system for on-off telegraph reception around
1928. This employed three antennas spaced about 1000 feet apart. The rectified
outputs of three separate receivers were combined across a common load
resistor and the voltage across the resistor keyed a local tone generator. As
long as the voltage was above a certain minimum from any receiver, a properly keyed tone signal was reproduced.

SOURCE: "Radio Receivers - past and present" by C. Buff. Proc. IRE (May
1962) p. 888.

SEE ALSO: "Diversity receiving system of RCA Communications Inc. for
radiotelegraphy". H. H. Beverage and H. O. Peterson. Proc. IRE, Vol. 19.
pp. 531-561. April 1931.

1928   FREQUENCY STANDARDS   J. W. Horton and W. A. Marrison   (U.S.A.)
        (Quartz Clocks)

        The tuning fork was developed to a point at which it gave a stability of
1 part in 10^7 per week and could have been improved still further. By this time,
however, the first quartz clock had been made by Horton and Marrison and it

seemed clear that quartz possessed many advantages. One fundemental advantage was the higher frequency of quartz vibrations. Frequencies of many mill-
ions of cycles per second were already being used for radio transmissions, and
it was not very convenient to measure them in terms of a standard having such
a low value as 1 kc.

SOURCE: "Frequency and time standards" L. Essen. Proc. IRE May 1962. p.1159

SEE ALSO: "Precision determination of frequency" J. W. Horton and W. A.
Marrison. Proc. IRE Vol.16. p.137. Feb.1928.

1928   PENTODE TUBE   Tellegen and   Holst (Holland)

        The substantial suppression of secondary emission in a tetrode
is not an easy matter, particularly where it is desired to operate with high
anode and screen potentials, and so by far the most common method of
suppressing secondary emission is by way of the inclusion of a suppressor
grid, between screen grid and anode, as in the pentode invented by Tellegen
and Holst of the Philips Company in Holland. The suppressor grid is
maintained at the filament potential. Pentodes, before 1939, had become
extremely popular for both high- and low-frequency amplification.

SOURCE: "Thermionic devices from the development of the triode up
to 1939" by Sir Edward Appleton. IEE Pub. Thermionic Valves 1904-1954.
IEE London 1955. p.23-24.

1929   MICROWAVE COMMUNICATION   A. G. Clavier (France)

        In 1920 Barkhausen positive-grid oscillator provided a means for the
efficient generation of 40-cm waves. This revived the interest in the centi-
meter waves. In 1929 Andre G. Clavier, then associated with Laboratoire
Central de Telecommunications in Paris, started an experimental project to
challenge the then accepted principle that wire or cable circuits should be used
in preference to radio whenever physically possible.
        In 1930 a link was started between two terminals in New Jersey using
10-ft. parabolic antennas. Just as testing started, the project was transferred back to France. On March 31st, 1931, Clavier and his associates demonstrated that microwave transmission provided a new order of economy,
quality, dependability and flexibility in communications over a 40-km path
between Calais and Dover. The circuit provided both telephone and tele-
printer service using 17.6-cm waves transmitted in a 4° beam by means of a
parabolic reflector 3 meters in diameter with a power output of a fraction of
a watt. Andre Clavier went on to establish the first commercial microwave
radio link in 1933 from Lympne, England, to St. Inglevert, France.

SOURCE: "Microwave Communications" by J. H. Vogelman. Proc. IRE (May
1962) p. 907.

1929   CYCLOTRON   E. O. Laurence (U.S.A.)

        Laurence used a curved path for the particles, so that the particles could circulate continuously, travelling long distances in a relatively small volume and using the same accelerating system over and over again. An electrically charged particle entering a magnetic field directed at right angles to the motion of the particle, proceeds to move in a circle with constant speed; as the particle speed is increased, the radius of the circle in which the particle moves also increases. Further acceleration occurs at each revolution.

SOURCE: "The Sources of Invention" by J. Jewkes, D. Sawers and R. Stellerman. MacMillan, London 1958. pp. 290/1

SEE ALSO: "Atomic Slingshot" by Howard Blakeslee. Science Digest, April, 1949.

ALSO: "Maestro of the Atom" by L. A. Schuler. Scientific American. August 1940.

1930   VAN DE GRAAF ACCELERATOR   V. de Graaf   (U.S.A.)

        For nuclear structure research, constant-potential accelerators use the electrostatic belt generator invented by Van de Graaf. About 5.5 million volts were insulated in air between two large generators in 1930 (equipment now in the Boston Museum of Science).

SOURCE: The Encyclopaedia of Physics (2nd Edition) Editor R. M. Besancon. Van Nostrand. New York 1974. p. 13.

SEE ALSO: R. J. Van de Graaf, J. G. Trump and W. W. Bruechner "Electrostatic generators for the acceleration of charged particles" Rept. Progr. Phys. 11 p.1. (1948)

1930   TRANSISTOR (MOSFET)   J. Lilienfeld (Germany)

        A 1930 patent was issued to Julius Lilienfeld of the University of Leipzig for a device that could be compared to today's MOSFET, or insulated-gate field-effect transistor. The device was reported to provide a means of obtaining amplification in a thin film of copper sulfide. However, a working device was probably never built, since the low mobility of holes in the material and other factors would seem to preclude any amplification.

SOURCE: "Solid State Devices" - "Electronic Design" 24. Nov. 23, 1972. p. 72.

1930's RADIOPHONIC SOUND - MUSIC   P. Grainger (Aus.)

        These techniques first came into real use during the 1950s with the maturation of the magnetic tape recorder although as long ago as the 1930s Percy Grainger, the Australian composer of "Country Gardens" fame, had produced a brief composition based on pure frequencies for the Theremin, an early electronic sound generator. The beginnings were however with musique concrete pioneered in Europe, although as the name suggests the "music" was made through the manipulation of pre-recorded natural sounds and was in fact orchestrated noise. With further study it became apparent that if more "musical" sounds were used as the raw material, i.e. sounds with a more ordered harmonic structure, greater malleability was achieved as the timbre changes encountered during pitch changes, due to differing tape speeds on playback, still bore some audible relationship to each other.

SOURCE: "Electronics Engineer's Reference Book" Newnes-Butterworth London (1976) Chap.17. p.17 - 16.

1931   FIXED RESISTOR (Oxide Film)   J.T.Littleton (U.S.A.)

The seed for this important contribution was provided by Littleton (1931) who developed an iridized, conducting tin-oxide coating for glass insulators. Its resistivity was sufficiently low to equalize potential across the insulator, thereby reducing corona effect, but too high for use in conventional resistors. Mochel modified this film by the addition of antimony oxide which stabilized its electrical properties. By varying the tin-antimony proportions, negative or positive temperature coefficients are obtained.

SOURCE: "Resistors - a survey of the evolution of the field" J.Marsten. Proc.IRE (May 1962) p.922.

SEE ALSO: J.T.Littleton. U.S.Patent No: 2,118,795 (1931)

ALSO: J.M.Mochel. U.S.Patent No:2,564,707 (1947) Reissue 25,556.

1931   STEREOPHONIC SOUND   A.D.Blumlein   (U.K.)
        REPRODUCTION   and Bell Labs   (U.S.A.)

Stereophonic reproduction per se was pioneered almost simultaneously by Blumlein in Great Britain and at the Bell Telephone Laboratories. Blumlein's contributions are presumed to be described in his patents. He showed a complete system applicable to sound-on-disc motion pictures, including microphone arrays utilizing bidirectional as well as omnidirectional microphones, transmission circuits, and disc recording systems utilizing simultaneous lateral and vertical recording. Economic difficulties are believed to have prevented completion and commercial exploitation of these systems.

The recognised early systems approach to large audience stereophonic reproduction was a public demonstration of Bell Telephone Laboratories equipment under the guidance of Dr.Harvey Fletcher on April 27, 1933. The Philadelphia Orchestra was in the Academy of Music in Philadelphia and it was reproduced in Constitution Hall, Washington, D.C.

SOURCE: "The history of stereophonic sound reproduction" by J.K.Hilliard Proc.IRE (May 1962) p.776.

SEE ALSO: A.D.Blumlein. Brit.Patent No:394,325. Dec.14,1931. also U.S. Patent No:2,093,540.

ALSO: "Perfect transmission and reproduction of symphonic music in auditory perspective" F.B.Jowett, et al. Bell Telephone Quart. Vol.12. July 1933. page 150.

1931   <u>C</u>OMPUTERS <u>(Differential Analyser)</u>   V. Bush   (U.S.A.)

Early analogue computer for solving differential equations.

<u>SOURCE</u>: "The differential analyser - a new machine for solving differential
equations" by V. Bush. Journal of the Franklin Institute (Vol. 212) 1931. p. 477

<u>SEE ALSO</u>: "The computer from Pascal to von Neumann" by H. H. Goldstine.
Princeton Univ. Press 1969. p. 88.

1931   <u>RELIABILITY - QUALITY CONTROL CHARTS</u>   W. A. Shewhart   (U.S.A.)

After languishing in libraries for several years, the work of Dodge and
Romig in acceptance sampling and the work of Shewhart on control charts
finally was brought to light during World War II through the nationwide training
programmes sponsored by the Office of Production Research and Development
of the War Production Board.

Although the underlying concepts were developed by scientific investigators and statisticians in the preceding decades, the genius of Dodge, Romig
and Shewhart lay in their recognition of basic principles as an aid to solving
practical problems, and their ability to recognize and formulate a systematic
approach.

<u>SOURCE</u>: "Treating real data with respect" by J. A. Henry. Quality Progress.
(March 1976) p. 18.

<u>SEE ALSO</u>:   "Economic Control of Quality of Manufactured Product" by
W. A. Shewhart. (1931) D. Van Nostrand Co. Inc., New York N. Y.

1932   <u>TRANSMISSION ELECTRON</u>   M. Knoll & E. Ruska (Germany)
        <u>MICROSCOPE</u>

The first electron microscope was built at the Technical University
of Berlin early in 1931. It had two electro-magnetic lenses in series and
achieved a modest magnification of 17. Improvements were made later. A
condenser lens was added and an iron shield with a narrow gap built round the
magnetic lens. Ruska in 1934 was able to demonstrate a magnification of
12.000.

<u>SOURCE</u>: "The Encyclopaedia of Physics" (2nd Edition) Editor R. M. Besancon
Von Nostrand. New York 1974. p. 275.

<u>SEE ALSO</u>: E. Ruska "Uber Fortschritte im Bau und in der Leistung des
Magnetischen Elektronemikroskops" Z. F. Physik 87. (9 & 10) 1934. p. 580.

ALSO: "Origin of the Electron Microscope" by M. M. Freundlich. Science 142.
(3589) 1963. p. 185.

1932   <u>CIRCUITRY (Energy Conserving</u>   A. D. Blumlein (U.K.)
         <u>Scanning Circuit)</u>

As with most of the diagrams, Figure 1(a) is taken from the
Patent Specification, and shows the basic features of the line scan circuit
which is now universal in television receivers, although not brought into

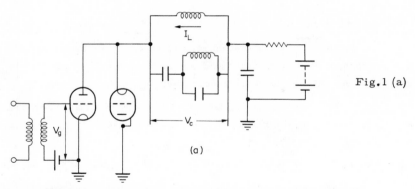

Fig.1 (a)

(a)

common use till 1946. Figure (1)b shows the method of operation, involving three separate regimes during the cycle.

It is now such a well-known circuit that it will not be described in detail, but it is interesting to compare it with the single LC circuit which is all that is necessary for a sinusoidal waveform whereas to handle the saw-tooth waveform it is necessary to provide also the switches in the form of the valve and diode as shown. The element of symmetry mentioned earlier can be seen here.

SOURCE: "The Work of Alan Blumlein" British Kinematography Sound and Television." Vol. 50. No. 7. July 1968. p. 209.

SEE ALSO: British Patent Specification No: 400, 976 (1932)

(b)

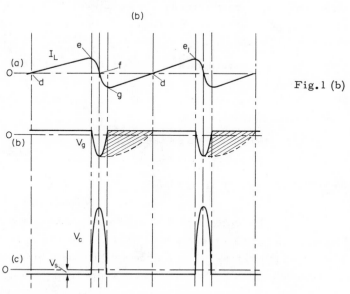

Fig.1 (b)

1933   "IGNITRON" (Mercury-arc Rectifier)   Westinghouse (U.S.A.)

In 1933 the Westinghouse Company announced its Ignitron. Its potential value was at once recognised and an active developmental programme soon commercialised it extensively. Progress in making and applying Ignitrons was rapid. By the end of 1934 a welding control unit using glass Ignitrons was installed in a customer's shop.

SOURCE: "Early history of industrial-electronics" W. C. White. Proc. IRE May 1962. p. 1133.

1933   RADIO ASTRONOMY   K.G.Jansky   (U.S.A.)

       While looking for the sources of static in overseas radio signals,
K.G.Jansky in 1933 discovered radio energy coming from the stars - thus
launching the science of radio astronomy.

SOURCE: "Mission Communications - the story of Bell Laboratories" by
Prescott C.Mabon. Published by Bell Laboratories Inc., Murray Hill,
New Jersey U.S.A.(1975) p. 170.

1933   HARD VALVE TIME BASE CIRCUIT   O.S.Puckle   (U.K.)

       In 1933 O.S.Puckle developed a time base which employs a variation
of the multivibrator as a condenser charging medium. This raised the maximum repetition frequency, as compared with that obtainable from a thyratron
time lease, from about 40 Kc/s up to a maximum of about 1 Mc/s.

SOURCE: "Time Bases" by O.S.Puckle.   Chapman & Hall, London, 1944.
p. 30.

SEE ALSO: "A time base employing hard valves" O.S.Puckle. British Patent
419198, also Journal of the Television Society Vol. 2. (1936) p. 147.

1933   FREQUENCY-MODULATION   E.H.Armstrong   (U.S.A.)

       The credit for promoting FM as a broadcast service goes to E.H.
Armstrong. For years he had been seeking a way to reduce static, and
finally he turned his attention to FM. Toward the end of 1933 he had perfected
a system of wide-band frequency modulation which seemed to overcome natural
and many forms of man-made static. In this system the carrier was frequency
modulated ±75 kc by audio components up to 15 kc.

SOURCE: "The development of the art of radio receiving from the early 1920's
to the present" by W.O.Swinyard. Proc.IRE (May 1962) p.797.
SEE ALSO: "Frequency modulation" by S.W.Seeley. RCA Rev. Vol. 5.
(April 1941) p. 468.

ALSO: "Selected papers on frequency modulation." Edited by J.Klapper. Dover
Publications Ltd., New York (1970)

1934   FREQUENCY STANDARDS (Atomic Clocks)   C.E. Cleeton
                                                   (U.S.A.)
                                    and N.A. Williams

       In 1934 Cleeton and Williams at Michigan University excited a spectral
line of ammonia at a frequency of 23, 870 Mcs by a source of radio waves generated in the laboratory. The source used by them for exciting the transitions was
a magnetron which generated a fairly wide band of frequencies and the ammonia
was at atmospheric pressure, at which only a very broad resonance effect is
observed. The width was mainly due to the effect of collisions and this can be
reduced by reducing the pressure.

SOURCE: "Frequency and time standards" L.Essen. Proc.IRE May 1962. p.1161

SEE ALSO: "Electromagnetic waves of 1.1 cm wavelength and the absorption
spectrum of ammonia" C.E.Cleeton and N.A.Williams. Phys.Rev. Vol.45.
p.234-237 February 1934.

1934   <u>LIQUID-CRYSTALS</u>   J. Dreyer (U.K.)

      Although the liquid-crystal state was first noted in 1889, it was not until around 1934 that serious consideration was given to these electro-optical devices, in the Marconi laboratories in England. John Dreyer found that their orderly molecular arrangement could be used to orient dye molecules for making polarisers - a method still used even though his work was done in the 1940s and patented in 1950. The present explosion in liquid-crystal research began in the 1960s when the RCA laboratories in the United States began to investigate them. Its course of development can be charted by counting the US and British patents that have been granted - one each year in 1936, 1946, 1950, 1951, 1963, 1965 and 1967. Then, suddenly, seven in 1968, 11 in 1969, at least two in 1970 and more than 11 in 1971. Few companies claim as long-lived an association with the subject as Marconi and RCA: most have been in the field for two years or less.

SOURCE: "The fluid state of liquid-crystals" by M. Tobias. New Scientist. 14 December, 1972. p. 651.

1935   <u>MULTIPLIER PHOTOTUBES</u>   Zworykin, Morton & Malter (U.S.A.)

      One of the most important by-products of television research is the multiplier phototube. When electroncs of one or several hundred electron volts energy impinge on a suitably prepared conducting surface, they eject 4 to 10 low-velocity electroncs, multiplying the initial current by a corresponding factor of 4 to 10. Repetition of this process leads to current multiplication by an arbitrarily high factor, practically without the addition of amplification noise. If the initial current is derived from a photocathode, the tube output reflects the variation of the light incident on the photocathode with a precision which depends only on the quantum efficiency of the cathode; with proper design, the dispersion in the transit time of the electrons from the cathode to the final collector can be held to quantities of the order of 10^{-10} second.

      In the earliest effective multipliers (Zworykin, Morton and Malter 1935) the electrons were guided from dynode to dynode along an approximately cycloidal path by crossed electric and magnetic fields. Purely electrostatic focusing and acceleration systems were developed subsequently by Zworykin and Rajchman and Rajchman and Synder as well as by Larson and Salinger. These may be regarded as the prototypes of present-day multiplier phototubes of RCA and DuMont. The venetian-blind design utilized in the image orthicon is also employed by EMI and RCA for multiplier phototubes. Finally, the early and very simple screen multiplier of Weiss (1936) does without focusing altogether, at the expense of materially lowered multiplication efficiency.

SOURCE: "Beam-deflection and photo devices" by K. Schlesinger and E. G. Ramberg. Proc. IRE (May 1962) p. 1001/2.

SEE ALSO: "The secondary emission multiplier - a new electronic device" by V. K. Zworykin, G. A. Morton and L. Malter. Proc. IRE Vol. 24 (March 1936) p. 351.

ALSO: "The electrostatic electron multiplier" by - and J. A. Rajchman. Proc. IRE Vol. 27. (Sept. 1939) p. 558.

ALSO: "Photocell multiplier tube" by C.C.Larson & H.Salinger Rev, Sci.
Instr. Vol.11 (July 1940) p.226.

ALSO: "On secondary emission multipliers" by G. Weiss. Z. Tech. Physik
Vol.17. (Dec.1936) p.623.

1935   TRAVELLING WAVE MICROWAVE   A. and O. Heil (Germany)
OSCILLATOR (early Magnetron)

      Studies of the classical triode valve in which the anode current is
controlled by the grid had shown that a fundemental difficulty for the highest
frequencies was the excess grid control power needed due to electron inertia.
In 1935 proposals were made by Arsenjewa Heil and O. Heil for avoiding this
limitation and also of avoiding the power dissipation limit of very high frequency circuits. These proposals were of particular importance since for the
first time, a new mechanism specially suited for the generation of very high
frequencies was suggested.

SOURCE: "Microwave valves: A survey of evolution, principles of operation
and basic characteristics" by C.H. Dix and W.E. Willshaw. Journal Brit. IRE
August, 1960. p.580.

SEE ALSO: "Eine neue methode zur erzeugung kurzer, ungedampfter,
elektromagnetischer Wellen grosser Intensitat" by A. Arsenjewa Heil and
O. Heil. Z.f. Phys. 95. p.752. 1935.

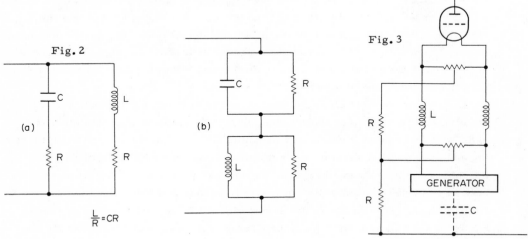

1935   CIRCUITRY (Constant Resistance   A.D. Blumlein (U.K.)
Capacity Stand-off Circuit)

      Figures 2(a) and (b) show two versions of the basis of this
invention, namely two-terminal arrangements of an inductor L, a
capacitor C, and two equal resistors R, having the property that the
impedance measured between the two terminals is purely resistive, of
value R, at all frequencies, provided $L/C = R^2$. This property was known,

but Blumlein adapted the circuit (particularly Figure 2(b)) as a means of removing from critical points in a circuit (e.g. a wide-band amplifier) the stray capacity to earth of, for example, floating power supplies. The example shown in Figure 3 is the application of the idea to the filament supply for the cathode follower output valve of the vision modulator for the original Alexandra Palace transmitter. The 'hardware' of this is preserved in the Science Museum.

SOURCE: "The Work of Alan Blumlein" British Kinematography Sound and Television. Vol. 50. No. 7. July 1968. p. 209.

SEE ALSO: British Patent Specification No: 462, 530 (1935)

1935   TRANSISTOR (field effect)   O. Heil   (Germany)

        In 1935 Oskar Heil of Berlin obtained a British patent on "Improvements in or Relating to Electrical Amplifiers and Other Control Arrangements and Devices". Figure 4 is the inventor's original illustration describing his device. The light area marked 3 is described as a thin layer of a semiconductor such as tellurium, iodine, cuprous oxide, or vanadium pentoxide; 1 and 2 designate ohmic contacts to the semiconductor. A thin metallic layer marked 6 immediately adjacent to but insulated from the semiconductor layer serves as control electrode. Heil describes how a signal on the control electrode modulates the resistance of the semi-conductor layer so that an amplified signal may be observed by means of the current meter 5. Using today's experience and language, one might describe this device as a unipolar field-effect transistor with insulated gate.

SOURCE: "The field-effect transistor - an old device with new promise" by J. T. Wallmark. IEEE Spectrum. (March 1964) p. 183.

SEE ALSO: British Patent No: 439, 457.

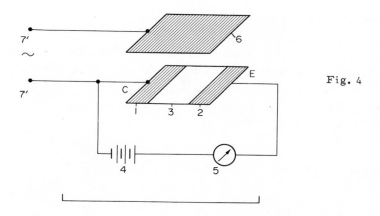

Fig. 4

1935   SCANNING ELECTRON MICROSCOPE   M. Knoll (Germany)

   The scanning electron microscope is a recent addition to the tools
available to the research worker. Postulated in the 1930s by Knoll and von
Ardenne, serious design only started in 1948 under Oatley in Cambridge,
resulting in commercial production in 1965. Such is the universal applicability of the technique that at the end of 1973 there were probably 2000 units
engaged in active research worldwide. The chief property of the s.e.m.
is its ability to image rough surfaces non-destructively with a minimum of
sample preparation.

SOURCE: "Electronics Engineer's Reference Book" Newnes-Butterworth
London (1976) Chap. 14. p. 14 - 79

SEE ALSO: "Aufladepotential und Sekundar-emission elektronbestrahlter
Oberflachen" by M. Knoll. Z. Techn. Phys. 2. 467 (1935)

ALSO: "Das Elektronen raster mikroskop" by M. vonArdenne. Z. Techn.
Phys. 19. 407-416 (1938)

1936   VOCODER   Bell Laboratories (U.S.A.)

   In 1936, Bell Laboratories developed the voice coder, or vocoder, for
analyzing the pitch and energy content of speech waves. With later developments, vocoder output was digitized, encrypted and the digital signal transmitted within a voice channel. The vocoder has been used since World War II
by the U.S. Government for secure communications.

SOURCE: "Mission Communications - the story of Bell Laboratories" by
Prescott C. Mabon. Published by Bell Laboratories Inc., Murray Hill,
New Jersey U.S.A. (1975) p. 171.

1936   COLD CATHODE TRIGGER TUBE   Bell Laboratories (U.S.A.)

   Bell Telephone Laboratories announced the first cold cathode
trigger tube in 1936. Using an activated cathode, the 313-A set the pattern
for a number of trigger tubes made in Europe during the next two decades.
The activated cathode led to anode maintaining and critical trigger voltages
each of the order of 70 V.

SOURCE: "A Survey of Cold Cathode Discharge Tubes" by D. M. Neale
The Radio and Electronic Engineer. February 1964. p. 87.

SEE ALSO: "The 313-A vacuum tube" by S. B. Ingram. Bell Lab. Rec.,
p. 114-6. December 1936.

1936   WAVEGUIDES   J. R. Carson, S. P. Meade
                             S. A. Schelkunoff and   Bell Laboratories
                             G. C. Southworth   (U.S.A.)

   In 1936, from the Bell Telephone Laboratories, Carson, Meade and
Schelkunoff published their mathematical theory on "Hyper-Frequency Wave-

Guides" while G. C. Southworth published his experimental results. These
papers provided the basis for the TE01 mode cylindrical waveguide. In that
same year W. L. Barrow of M. I. T. published his work on the "Transmission
of Electromagnetic Waves in Hollow Tubes of Metal". Before 1934 Southworth
had transmitted telegraph and telephone signals at 15-cm wavelengths in a 5-in
diameter hollow metal pipe 875 ft. long with relatively small attenuation.

SOURCE: "Microwave Communications" by J. H. Vogelman. Proc. IRE (May
1962) p. 907.

SEE ALSO: (1) "Hyperfrequency waveguides - mathematical theory" by
J. R. Carson, S. P. Meade and S. A. Schelkunoff. Bell Sys. Tech. J. Vol. 15.
p. 310-333 April 1936.

          (2) "Hyperfrequency waveguides - general considerations and
experimental results" G. C. Southworth. Bell Sys. Tech. J. Vol. 15. p. 284-309
April 1936.

1936   CIRCUITRY (Long-Tailed Pair)   A. D. Blumlein (U. K.)

          This now familiar and much used circuit (Figure 5 (a) and (b)) was
first needed in the amplifiers for the original video cable between points in
Central London and Alexandra Palace. The cable was not the now familiar
co-axial type but a shielded pair, and the problem was to obtain the 'push-
pull' signal uncontaminated by 'push-push' interference pick-up. In telephone
practice a transformer serves this purpose, but transformers to handle the
video frequency range were not then available.

          The name of the circuit invented to do the job is the name given to
it by Blumlein, and is so descriptive that it has stuck.

Fig. 5 (a)

Fig. 5 (b)

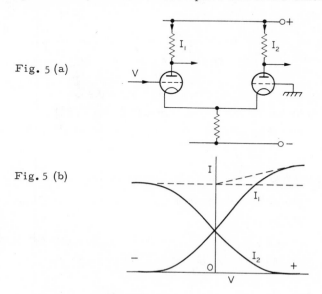

SOURCE: "The Work of Alan Blumlein" British Kinematography Sound
and Television. Vol. 50. No. 7. July 1968. p. 209.

SEE ALSO: British Patent Specification No: 482, 740 (1936)

1937   <u>XEROGRAPHY</u>   Chester Carlson (U.S.A.)

       An individual inventor, Chester Carlson, conceived the idea of
Xerography. This is a new photographic process which in a relatively short
time has found numerous industrial applications. It is completely dry and is
based entirely upon principles of photoconductivity and electrostatics. The
process:

> "employs a plate which consists of a thin photoconductive
> coating on a metallic sheet. This coating can be electrically
> charged in the dark and will hold this charge until exposed to
> light. Thus an electrostatic image can be produced on the
> plate by exposing the plate to an optical image. When the plate
> is dusted with powder particles, the electrostatic image is
> transformed into a powder image which can be transferred to
> paper and fixed by fusing."

       The development of Xerography was turned over to Roland M.
Schaffert, a Battelle research physicist with some previous experience in
printing. For a year he worked alone but after the war Battelle assigned a
few assistants to help him. By the latter part of 1946 two important
developments were completed: a high-vacuum technique for coating plates
with selenium; and a corona discharge wire, both for applying the original
electrostatic charge to the plate and for transferring powder from the plate
to the paper. The most significant contribution was the discovery of a
method to keep the image background from being filled with stray powder.
Thus Battelle improved Xerography to the point where industry became
interested.

SOURCE: "The Sources of Invention" by J.Jewkes, D.Sawers and
R.Stillerman. MacMillan & Co. London. 1958. p. 405 & 408.

SEE ALSO: "Printing with Powders" Fortune. June 1949.
                 "Xerography - From Fable to Fact" by W.T.Reid.
                 "Developments in Xerography" by R.M.Schaffert. The Penrose
                 Annual, 1954.

1938   <u>COMPUTERS - (Information Theory)</u>   C.E.Shannon   (U.S.A.)

       Shows the analysis of complicated circuits for switching could be
effected by the use of Boolean algebra.

SOURCE: "A symbolic analysis of relay and switching circuits" C.E.Shannon.
Trans. A.I.E.E. 1938. Vol.57. p.713.

1939   <u>KLYSTRON</u>   W.C.Hahn and Varian Bros.   (U.S.A.)

       Perhaps the first great step in understanding the phenomena in micro-
wave tubes came with the invention of the klystron. Bruche and Recknagel
discussed "phase focusing" in 1938 and the work of the Varians, Webster's
theoretical treatment of the klystron and the work of Hahn and Metcalf were
published in 1939.

With the klystron came a well-thought-out theory of its operation, the concept of velocity modulation, and a full appreciation of the value of microwave resonators.

SOURCE: "History of the microwave tube art" by J.R.Pierce. Proc.IRE (May 1962) p.979.

SEE ALSO: "High frequency oscillator and amplifier" by R.H. and S.F.Varian J.App.Phys. Vol.10 (May 1939) p.321.

1939   BELL TELEPHONE LABS "COMPLEX   G.Stibitz et al (U.S.A.)
        COMPUTER"

It is perhaps a little surprising that it was not until 1937 that Bell Telephone Laboratories investigated the design of calculating devices, although Andrews has stated that from about 1925 the possibility of using relay circuit techniques for such purposes was well accepted there. However, in 1937 George Stibitz started to experiment with relays, and drew up circuit designs for addition, multiplication and division. At first he concentrated on binary arithmetic, together with automatic decimal-binary and binary-decimal conversion, but later turned his attention to a binary-coded decimal number representation. The project became an official one when, prompted by T.C.Fry, Stibitz started to design a calculator capable of multiplying and dividing complex numbers, which was intended to fill a very practical need, namely to facilitate the solution of problems in the design of filter networks, and so started the very important Bell Telephone Laboratories Series of Relay Computers.

In November 1938, S.B.Williams took over responsibility for the machine's development and together with Stibitz refined the design of the calculator, whose construction was started in April and completed in October of 1939. The calculator, which became known as the "Complex Number Computer", often shortened to "Complex Computer" and as other calculators were built, the "Model 1" began routine operation in January 1940. Within a short time it was modified so as to provide facilities for the addition and subtraction of complex numbers, and was provided with a second, and then a third teletype control, situated in remote locations. It remained in daily use at Bell Laboratories until 1949.

SOURCE: "The Origins of Digital Computers" Edited by B.Randell. Springer-Verlag, Berlin. (1973) p.238.

SEE ALSO: "Computer" by G.R.Stibitz - "The Origins of Digital Computers" Edited by B.Randell. Springer-Verlag, Berlin. p.241.

1939   FREQUENCY STANDARDS (Caesium Beam)   I.I.Rabi (U.S.A.)

The difficulties of bandwidth and low intensity are most easily overcome by using the atomic beam magnetic resonance method developed at Columbia University by Rabi and his co-workers. In this method which can be used with atoms possessing a magnetic dipole moment, a beam of atoms passes to a detector through a system of magnets and a region of field alternating at the Bohr frequency. The magnets have a nonuniform field and deflect the atoms in one direction or the other according to which of the two energy levels they are in.

When the frequency of the RF field is exactly equal to the Bohr frequency and is of the right amplitude transitions are induced and the deflections in the second magnet B are the opposite from those in the first magnet. A, and the atoms are thus focused on the detector.

SOURCE: "Frequency and time standards" L. Essen. Proc. IRE May 1962. p. 1162

1939   MAGNETRON   J. T. Randall and H. A. H. Boot   (U. K.)

In the autumn of 1939 the Admiralty asked Professor M. L. Oliphant and the physics department of the University of Birmingham to develop a high-power micro-wave transmitter. The majority of the scientists in the laboratory concentrated on the klystron, described by its inventors, R. H. and S. F. Varian of Stanford University, California, in 1939, which used for the first time closed resonators, described by W. W. Hansen, also of Stanford, in 1938, for the production of high-frequency power. J. T. Randall and H. A. H. Boot, struck by the difficulty of getting enough power from the klystron, considered instead applying the resonator principle to the magnetron, which had been invented by A. W. Hull of the American General Electric firm in 1921 but which, in its conventional form, lacked the properties they were seeking. The result was the cavity magnetron, which proved to be the needed generator, producing high powers on centimetre wave-lengths.

SOURCE: "The Sources of Invention" by J. Jewkes, D. Sawers and R. Stillerman MacMillan & Co. London 1958. p. 348.

1939   COMPUTERS (Digital)   H. H. Aitken (U. S. A.) and I. B. M.

Utilizing twentieth-century advances in mechanical and electrical engineering, the Automatic Sequence Controlled Calculator, or Mark I, brought Babbage's ideas into being, giving concrete existence to much more at the same time. The Mark I, an electromechanical calculator 51 feet long and 8 feet high, was built by the International Business Machines Corporation between 1939 and 1944. It could perform any specified sequence of five fundamental operations, addition, subtraction, multiplication, division and reference to tables of previously computed results. The operation of the entire calculator was governed by an automatic sequence mechanism. The machine consisted of 60 registers for constants, 72 adding storage registers, a central multiplying and dividing unit, means of computing the elementary transcendental functions $\log_{10} x$, 10^x and $\sin x$, and three interpolators reading functions coded in perforated paper tapes. The input was in the form of punched cards and switch positions. The output was either punched into cards or printed by electric typewriters.

SOURCE: Serrell, Astrahan, Patterson and Pyrne "The evolution of computing machines and systems" Proc. IRE May 1962. p. 1043.

SEE ALSO: "The computer from Pascal to von Neumann" by H. H. Goldstine. Princeton Univ. Press 1972. p. 118.

ALSO: "Proposed automatic calculating machine" by H. H. Aitken. IEEE Spectrum. August 1964. p. 62.

1940   MICROELECTRONICS   Centralab (U.S.A.)
        (Thick film circuits)

       During World War II the Centralab Div. of Globe-Union Inc.,
developed a ceramic-based circuit for the National Bureau of Standards. This
"printed circuit" used screen-deposited resistor inks and silver pastes to
support the miniature circuits in an Army proximity fuse. The PC board that
followed stimulated manufacturers to develop components with radial leads
and tubular shapes.

SOURCE: "Solid State Devices - Packaging and Materials" R. L. Goldberg
"Electronic Design" 24. November 23, 1972. p. 127.

Proximity Fuze Circuits made in 1945.

1942   CIRCUITRY (Miller Integrator)   A. D. Blumlein (U.K.)

       Figure 6 shows this very well-known circuit. The name "Miller"
commemorates the man who pointed out (the usually deleterious effect) that
gain in a triode valve effectively increases the input capacitance by an amount
equal to the product of the gain and the stray grid/anode capacitance. Blumlein,
however, first made use of this by deliberately adding an external capacitor to
the stray capacitance. Although the date of this patent is 1942, the idea had
been used by Blumlein some years earlier, and in fact appears in a somewhat
elaborate form in his television frame scan circuit BPS 479, 113 (1936).

SOURCE: "The Work of Alan Blumlein" British Kinematography Sound
and Television. Vol. 50. No. 7. July 1968. p. 211.

SEE ALSO: British Patent Specification No: 580, 527 (1942)

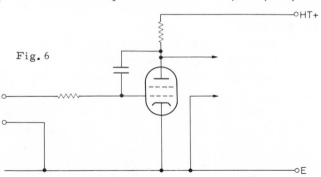

Fig. 6

1943   RELIABILITY (Sequential analysis)   A. Wald (U.S.A.)

        A major technical breakthrough occurred in the spring of 1943, when the noted mathematical statistician, Abraham Wald, devised his now celebrated basic theory of sequential analysis for analyzing U.S. war problems. Initial application of the theory in analyzing combat experience demonstrated its value in obtaining reliable conclusions from a minimum of information, swiftly and economically. This seemed like an ideal tool for use in quality control at this crucial period in our history when speed, precision, reliability and economy in production were of the essence. To give the new tool a thorough trial, the military released a multiple sampling plan based on Wald's sequential theory. This plan was first released to a limited number of strategic manufacturing firms for use in acceptance sampling. The early success of the plan and the subsequent widespread demands for its use resulted in removal of its "restricted" classification in 1945.

SOURCE: "The reliability and quality control field from its inception to the present" by C. M. Ryerson. Proc. IRE (May 1962) p. 1326.

1943   PRINTED WIRING   P. Eisler (U.K.)

        This invention relates to the manufacture of electric circuits - such, for instance, as that of a telephone switchboard - and circuit components - such as inductances, resistance, magnetic cores and their existing windings; and consists in producing them by the methods of the printing art or methods akin to them, methods, that is to say, by which the conductor of the circuit is brought into existence in its final form, or a development of that form upon a plane or other surface, instead of being first produced as a linear conductor and afterwards given its three-dimensional form.

        A typical instance of the invention comprises the steps of making a drawing of the electric or magnetic circuit, or of a development of it if it is of three dimensions: preparing from that drawing, by any of the well-known methods of the printing art, a printing surface; making an imprint by the aid of the printing surface; and from that imprint producing the conductor.

SOURCE: U.K. Patent No: 639, 178. February 2nd, 1943.

SEE ALSO: "Technology of Printed Circuit" by P. Eisler. Heywood, London (1959).

1943   TRAVELLING WAVE TUBE   R. Kompfner, A. W. Haeff and
                         J. R. Pierce   (U.S.A.)

        It remains, however, for Kompfner to take the decisive step of reasoned approach and effective experiment which gave us the travelling-wave tube and led to a host of related devices.

        Kompfner reasoned that an electromagnetic wave on a slow-wave structure, and especially a helix, should interact powerfully with a beam of electroncs if the wave velocity and the electron velocity were nearly the same. He built a tube and found a gain of around 10 db for a wave travelling in the same direction as the electrons. A mathematical analysis agreed with this performance.

The travelling-wave tube turned out to be a device which amplifies over unprecedently broad bands, and which can function over an astonishingly wide range of frequencies and powers. A travelling-wave tube with a helix circuit can amplify over a frequency range of more than an octave. More typically, bandwidth is limited to 500-1000 Mc by input and output couplers.

SOURCE: "History of the microwave tube art" by J.R.Pierce. Proc.IRE (May 1962) p.980.

SEE ALSO: "The travelling wave valve" by R.Kompfner. Wireless World Vol.52. (Nov.1946) p.369.

1943   COMPUTERS (ENIAC)   Moore School (U.S.A.)
(Electronic numerator, integrator
and computer)

The ENIAC was developed and built at the Moore School of Electrical Engineering of the University of Pennsylvania, baginning in 1942 and was completed in 1946. Its principal object was the computation of firing and ballistic tables for the Aberdeen Proving Ground of the U.S.Army Ordnance Corps. This computation required the integration of a simple system of ordinary differential equations involving arbitrary functions.

This equipment occupied a space 30 x 50 feet and contained 18000 vacuum tubes. The computing elements consisted largely of decade rings, flip-flops and pentode gates. The input-output system consisted of modified IBM card readers and punches.

SOURCE: Serrell, Astrahan, Patterson and Pyrne "The evolution of computing machines and systems" Proc. IRE May 1962. p.1044.

SEE ALSO: "The computer from Pascal to von Neumann" by H.H.Goldstine. Princeton Univ.Press 1972. p.117.

1944   RELIABILITY - SAMPLING   H.F.Dodge & H.G.Romig   (U.S.A.)
INSPECTION TABLES

After languishing in libraries for several years, the work of Dodge and Romig in acceptance sampling and the work of Shewhart on control charts finally was brought to light during World War II through the nationwide training programmes sponsored by the Office of Production Research and Development of the War Production Board.

Although the underlying concepts were developed by scientific investigators and statisticians in the preceding decades, the genius of Dodge, Romig and Shewhart lay in their recognition of basic principles as an aid to solving practical problems, and their ability to recognize and formulate a systematic approach.

SOURCE: "Treating real data with respect" by J.A.Henry. Quality Progress. (March 1976) p.18.

SEE ALSO: "Sampling Inspection Tables" by H.F.Dodge and H.G.Romig. 2nd ed. 1944. John Wiley & Sons, New York, N.Y.

1945   <u>COMPUTERS (Whirlwind)</u>   Massachusetts Institute of Technology. (U.S.A.)

        An assignment to build a real-time aircraft simulator was given in
1945 to the Digital Computer Laboratory of the Massachusetts Institute of
Technology, at that time a part of the Servomechanisms Laboratory of M.I.T.
Beginning in 1947, the major part of the effort was devoted to the design and
construction of the electronic digital computer known as 'Whirlwind I'. The
project was sponsored by the Office of Naval Research and the United States Air
Force. The machine was put in operation in March 1951.

        'Whirlwind I' was a parallel, synchronous, fixed-point computer
utilizing a number length of 15 binary digits plus sign (16 binary digits in all).
Physically, it was a large machine containing some 5000 vacuum tubes (mostly
single pentodes) and some 11,000 semiconductor diodes. It consisted of an
arithmetic "element" including three registers; a control element including
central control, storage control, arithmetic control and input-output control;
a program counter - a source of synchronizing pulses or master clock supply-
ing 2 megapulses per second to the arithmetic element and 1 megapulse per
second to the other circuits; an internal storage element or memory, terminal
equipment; and extensive test and marginal checking equipment.

SOURCE: Serrell, Astrahan, Patterson and Pyne "The evolution of computing
machines and systems" Proc. I RE May 1962 '.1047.

SEE ALSO: "The computer from Pascal to von Neumann" by H.H. Goldstine.
Princeton Univ. Press 1972. p. 212.

1945-   <u>POTTED CIRCUITS</u>   U.K. and U.S.A.)
1950
        The potting of electrical apparatus in wax or bitumen compounds was
carried out for many years, but it is only recently that plastics suitable for this
purpose have become available, in the form of cold-polymerizing casting resins.
There is no doubt that the small sub-unit is now as essential part of modern
electronic equipment, and potting techniques lend themselves to this construction,
provided that means are available to dissipate the heat developed. The resins
are relatively expensive, and mechanically and economically are not attractive
for castings exceeding a few inches in major dimensions.

        The casting resins are converted into rigid plastics by the addition of a
catalyst and accelerator, without the application of the considerable pressures
and temperatures normally associated with the polymerization of thermo-setting
resins.

SOURCE: "New constructional techniques" by G.W.A. Dummer and D.L.
Johnston. Electronic Engineering (Nov.1953) p.456.

SEE ALSO: "How plastics aid miniaturization of electrical assemblies"by
R.J. Bibbero and E.B. Chester. Mach. Des. (Oct.1951) p.127.

ALSO: "Potted Circuits - new development in miniaturization of equipment"
Wireless World, 57, 493 (1951)

ALSO: "Cast resin embedments of circuit sub-units and components" Elect.
Mnfg. 48, 103 (1951).

1945   <u>COMPUTERS (Theory)</u>   Von Neumann (U.S.A.)

      Basic design of the electronic computer project of the Institute of Advanced Study incorporating ideas underlying essentially all modern machines.

<u>REFERENCE</u>: "Memorandum on the program of the High Speed Computer" by von Neumann. 8th November, 1945.

<u>SEE ALSO</u>: "The computer from Pascal to von Neumann" by H.H. Goldstine. Princeton Univ. Press 1972. p. 255.

1945   <u>COMMUNICATION (Satellite)</u>   A.C. Clarke (U.K.)

      Early interest in space was concentrated upon the propulsion aspect, and the forthcoming marriage of space and electronics had to await the publication of A.C. Clarke's paper on communications satellites in 1945. The use of a satellite S above the radio horizons of both A and B permitted microwave transmission from A to S, and from S to B, thus bridging the oceans by microwave link. Clarke's paper further pointed out that at an orbital altitude intermediate between a 90-minute SPUTNIK and the 28-day orbit of the Moon there was an orbit taking one day, so that an easterly-launched satellite above the Equator would give the radio engineer an imaginary mast 22 300 miles high on which to place his aerials.

<u>SOURCE</u>:   "Electronics in Space" by W.F. Hilton. The Radio and Electronic Engineer. Vol. 45. No. 10. (Oct. 1975) p. 623.

<u>SEE ALSO</u>: "Extra terrestial relays" by A.C. Clarke. Wireless World Vol. 51. No. 10. (Oct. 1945) p. 305.

1946   <u>COMPUTERS (CRT Storage)</u> F.C. Williams (U.K.)

Storage of pulses on the face of a CRT as a memory device.

<u>REFERENCE</u>: "A Storage System for use with Binary Digital Computers" by F. C. Williams and T. Kilburn. Proc. IEE Vol. 96. Pt. 2. No. 81. (1949) p. 183.

<u>SEE ALSO</u>: "The computer from Pascal to von Neumann" by H.H. Goldstine. Princeton Univ. Press 1972. p. 248.

1947   <u>CHIRP RADAR TECHNIQUES</u>   Bell Laboratories   (U.S.A.)

      The Chirp or pulse-compression technique for radar originated at Bell Laboratories in 1947. This technique, in which long, modulated pulses are transmitted and then compressed upon reception, permitted pulsed radar systems to have long range and high resolution while avoiding problems associated with generating and transmitting short pulses with high peak powers.

<u>SOURCE</u>: "Mission Communications - the story of Bell Laboratories" by Prescott C. Mabon. Published by Bell Laboratories Inc., Murray Hill, New Jersey U.S.A. (1975) p. 179.

1947   ECME (Electronic circuit making equipment)   J. A. Sargrove (U. K.)

      The process of John. A. Sargrove for the automatic assembly of electronic apparatus was the first modern approach to automatic operation in electronic manufacturing. In 1947 he built and operated a machine for the automatic production of two- and five-tube radio receivers.

      The first operation of Sargrove's machine was to prepare the $\frac{1}{4}$-in. molded-plastic plates by blasting with an abrasive grit to roughen both sides of the plates simultaneously. The plates were then triple-sprayed with zinc to form the conducting surface. The spraying machine consisted of eight nozzles arranged four to a side to allow simultaneous spraying of both sides of the plate once it was positioned. Materials to form resistance, capacitance and conductors were sprayed through stencils onto their proper positions on the plate.

SOURCE: "Electronic equipment design and construction" by G. W. A. Dummer, C. Brunetti and L. K. Lee. McGraw-Hill, New York (1961) p. 192-193.

SEE ALSO: "New methods of radio production" by J. A. Sargrove. J. Brit. IRE Vol. 7(1) Jan/Feb. 1947. p. 2.

ALSO: "Automatic Receiver Production" Wireless World. April, 1947.

Examples of radio sets produced by automatic assembly methods.
(Courtesy - Sargrove Electronics)

1947   COMPUTERS (EDVAC)   University of Pennsylvania   (U. S. A.)

      The Electronic Discrete Variable Automatic Computer, or EDVAC, was built at the Moore School (University of Pennsylvania) between 1947 and 1950 for the Ballistic Research Laboratory at the Aberdeen Proving Ground. It is a serial, synchronous machine in which all pulses are timed by a master clock operating at 1 mega-pulse per second. It contains some 5900 vacuum tubes, about 12000 semi-conductor diodes and utilizes the binary number system with a word length of 44 binary digits.

SOURCE: Serrell, Astrahan, Patterson and Pyne "The evolution of computing machines and systems" Proc. IRE May 1962. p. 1046.

SEE ALSO: "The computer from Pascal to von Neumann" by H. H. Goldstine Princeton Univ. Press 1972. p. 187.

1947   HIGH QUALITY AMPLIFIER CIRCUIT   D. T. N. Williamson   (U.K.)

      The requirements of such an amplifier may be listed as:-

(1) Negligible non-linear distortion up to the maximum rated output. (The term "non-linear distortion" includes the production of undesired harmonic frequencies and the intermodulation of component frequencies of the sound wave.) This requires that the dynamic output/input characteristic be linear within close limits up to maximum output at all frequencies within the audible range.

(2) (a) Linear frequency response within the audible frequency spectrum of 10-20,000 c/s.   (b) Constant power handling capacity for negligible non-linear distortion at any frequency within the audible frequency spectrum.

(3) Negligible phase-shift within the audible range. Although the phase relationship between the component frequencies of a complex steady-state sound does not appear to affect the audible quality of the sound, the same is not true of sounds of a transient nature, the quality of which may be profoundly altered by disturbance of the phase relationship between component frequencies.

(4) Good transient response. In addition to low phase and frequency distortion, other factors which are essential for the accurate reproduction of transient wave-forms are the elimination of changes in effective gain due to current and voltage cut-off in any stages, the utmost care in the design of iron-cored components, and the reduction of the number of such components to a minimum.

(5) Low output resistance. This requirement is concerned with the attainment of good frequency and transient response from the loudspeaker system by ensuring that it has adequate electrical damping.

(6) Adequate power reserve. The realistic reproduction of orchestral music in an average room requires peak power capabilities of the order of 15-20 watts when the electro-acoustic transducer is a baffle-loaded moving-coil loudspeaker system of normal efficiency.

SOURCE: "Design for a high-quality amplifier" Part I. by D. T. N. Williamson Wireless World (April 1947) p. 118

ALSO: "Design for a high--quality amplifier" Part 2. Wireless World (May 1947) p. 161.

1947   COMPUTERS (UNIVAC)   P. Eckert and J. Mauchly (U.S.A.)
Universal Automatic Computer

      The development of the Universal Automatic Computer, or UNIVAC was started about 1947 by Presper Eckert and John Mauchly who founded the Eckert-Mauchly Computer Corporation in December of that year. The first UNIVAC I was built for the U.S.A. Bureau of the Census and was put in operation in the spring of 1951. (The Eckert-Mauchly Corporation later became a subsidiary of Remington Rand, forming the organisation which is now the Remington Rand UNIVAC Division of the Sperry Rand Corporation).

      UNIVAC I was a direct descendent of the ENIAC and of the EDVAC in the development of which Eckert and Mauchly had both had an important part at the University of Pennsylvania. It was a serial, synchronous machine operating at a rate of 2.25 megapulses per second. It contained some 5000 tubes and several times as many semiconductor diodes in logic and clamp circuits. One hundred mercury delay lines provided 1000 twelve-decimal-digit words of internal storage. Twelve additional delay lines were used as input-output registers. Aside from console switches and an electric typewriter providing small amounts of information, the input-output medium was metal-base magnetic tape. Forty-eight UNIVAC I machines were built.

SOURCE: Serrell, Astrahan, Patterson and Pyne "The evolution of computing machines and systems" Proc. IRE May 1962. p. 1048/9.

SEE ALSO: "The computer from Pascal to von Neumann" by H. H. Goldstine. Princeton Univ. Press 1972. p. 246.

1948   TRANSISTORS - Single crystal   G. K. Teal & J. B. Little   (U.S.A.)
        fabrication - germanium.

        In the latter part of 1948 G. K. Teal and J. B. Little of BTL began experiments to grow germanium single crystals, selecting the pulling technique. They succeeded in growing large single crystals of germanium of high structural perfection. They also improved the impurity of the material by repeated recrystalization methods.
        At BTL Teal, working with M. Sparks, devised a unique method for preparing p-n junctions by modifying his crystal-pulling apparatus to allow controlled addition of impurities during crystal growth. Using ingots they prepared single crystals containing p-n junctions and soon afterwards n-p-n grown-junction transistors which had many of the properties predicted by Shockley.

SOURCE: "Contributions of materials technology to semiconductor devices" by R. L. Petritz Proc. IRE (May 1962) p. 1026.

ALSO: "Growth of germanium single crystals" G. K. Teal and J. B. Little. Phys. Rev. Vol. 78. p. 647. (June 1950)

ALSO: "Growth of silicon single crystals and of single crystal silicon p-n junctions" Phys. Rev. Vol. 87. p. 190 (July 1952)

1948   COMMUNICATION (Information theory)   C. E. Shannon (U.S.A.) *
        The term "Information Theory" is used in the current technical literature with many different senses. Historically it seems first to have been generally applied to describe the specific mathematical model of communication systems developed in 1948 by Shannon. In this pioneering paper, Shannon introduced a numerical measure, called by him and others entropy, of the randomness or uncertainty associated with a class of messages and showed that this quantity measures in a real sense the amount of communication facility needed to transmit with accuracy messages from the given class. He also showed (quite incidentally to his main argument) that this measure of uncertainty agreed in certain aspects with the common, vague intuitive notion of the "information content of a message". He accordingly used the words "information content" as a synonym for the precisely defined notion of entropy. As a result, his work and its immediate extensions became known as information theory.

SOURCE: "Information Theory" by B. McMillan and D. Slepian. Proc. IRE (May 1962) pp. 1151/2.

SEE ALSO: "The mathematical theory of communication" by C. E. Shannon and W. Weaver. University of Illinois Press, Urbana. 1949.

1948   COMPUTERS (SEAC)   Nat. Bureau of Standards   (U.S.A.)

The Standards Electronic Automatic Computer, SEAC, was built by the staff of the Electronic Computer Laboratory of the National Bureau of Standards. The design began in June 1948 and the machine was put in operation in May 1950. It was built under the sponsorship of the Office of the Air Comptroller, Department of the Air Force, principally to carry out mathematical investigations of techniques for solving large logistics programming problems.

SOURCE: Serrell, Astrahan, Patterson and Pyne "The evolution of computing machines and systems" Proc. IRE May 1962. p. 1046.

SEE ALSO: "The computer from Pascal to von Neumann" by H. H. Goldstine. Princeton Univ. Press 1972. p. 315.

1948   TRANSISTOR   Bardeen, Brattain and Shockley   (U.S.A.)

Immediately hostilities ceased in 1945 Shockley organised a group for research on the physics of solids. On testing out experimentally Shockley's ideas, it was discovered that the projected amplifier did not function as Shockley had predicted; something prevented the electric field from penetrating into the interior of the semi-conductor. John Bardeen, a theoretical physicist, formulated a theory concerning the nature of the surface of a semi-conductor which accounted for this lack of penetration of field, and also led to other predictions concerning the electrical properties of semi-conductor surfaces. Experiments were carried out to test the predictions of the theory. In one of these, Walter H. Brattain and R. B. Gibney observed that an electric field would penetrate into the interior if the field was applied through an electrolyte in contact with the surface. Bardeen proposed using an electrolyte in a modified form of Shockley's amplifier in which a suitably prepared small block of silicon was used. He believed that current flowing to a diode contact to the silicon block could be controlled by a voltage applied to an electrolyte surrounding the contact. In the earlier experiments testing Shockley's ideas, thin films with inferior electrical characteristics had been employed. Brattain tried Bardeen's suggested arrangement, and found the amplification as Bardeen had predicted, but the operation was limited to very low frequencies because of the electrolyte. Similar experiments involving germanium were successful, but the sign of the effect was opposite to that predicted. Brattain and Bardeen then conducted experiments in which a rectifying metal contact replaced the electrolyte and discovered that voltage applied to this contact could be used to control, to a small extent, the current flowing to the diode contact. Here again, however, the sign of the effect was opposite to the predicted one. Analysis of these unexpected results by the two scientists led them to the invention of the point-contact transistor, which operates on a completely different principle from the one first proposed. Current flowing to

one contact is controlled by current flowing from a second contact, rather
than by an externally applied electric field. Brattain and Bardeen used
extremely simple equipment, the most expensive piece of apparatus being an
oscilloscope.

The Bell Telephone Laboratories announced the invention in
June 1948, an, since then, development work has proceeded rapidly. The
first point-contact transistor had several limitations: it was noisy, it could
not control high amounts of power and it had a limited applicability.
Shockley had meanwhile conceived the idea of the junction transistor which
was free of many of these defects and most of the transistors now made are
of the junction type.

SOURCE: "The Sources of Invention" by J. Jewkes, D. Sawers and
R. Stillerman.   MacMillan & Co. London 1958. p. 400.

SEE ALSO: "The First Five Years of the Transistor" by Mervin Kelly.
Bell Telephone Magazine. Summer 1953.

1948   FILM SOUND RECORDING   RCA and others (U.S.A.)
        (Magnetic Film)

Although magnetic recording was one of the oldest methods known, it
was not until World War II that this form of recording came into its own.
During this period there was developed in Germany a fine grain, low-noise,
magnetic oxide, and a process for uniformly coating it on a thin flexible base
$\frac{1}{4}$ inch in width. Use of this new tape in properly designed recorders and repro-
ducers resulted in sound quality which was higher than had previously been
obtained from either the film or the disk method. Immediately after the war
some of the German recorders were demonstrated in this country, and the
potential impact of magnetic recording on the motion-picture industry was
quickly recognised.

Early in 1948, oxide coated 35 mm. film became available for use in
motion-picture sound recording. It was then possible to convert photographic
sound recorders to combination units càpable of recording either magnetic or
photographic sound. Many recorders were converted as quickly as possible and
were tried in motion-picture sound studios for original "takes". The tests of
magnetic recording in the studios were immediately successful, not only
because of its high-quality and large dynamic range, but also because magnetic
film provided a more flexible and a more economical means of recording. Since
re-recording of all original "takes" to a composite photographic negative was
already the accepted practice in the industry, little inconvenience was caused by
the change.

SOURCE: "Film Recording and Reproduction" by M. C. Batsel and G. L.
Dimmick. Proc. IRE (May 1962) p. 749.

1948   HOLOGRAPHY   D. Gabor (U. K.)

      With holography, one records not the optically formed image of an object but the object wave itself. This wave is recorded (usually on photograohic film) in such a way that a subsequent illumination of this record called a "hologram" reconstructs the original object wave. A visual observation of this reconstructed wavefront then yields a view of the object which is practically indiscernible from the original, including three dimensional parallax effects.

SOURCE: The Encyclopaedia of Physics (2nd Edition) Editor R. M. Besancon Van Nostrand. New York 1974. p. 426.

SEE ALSO: "Optical Holography" by R. J. Collier, C. B. Burckhardt and L. H. Lin. Academic Press. New York 1971.

1948   EDSAC (Electronic Delay Storage
Automatic Calculator)   M. V. Wilkes   (U. K.)

      The EDSAC (electronic delay storage automatic calculator) is a serial electronic calculating machine working in the scale of two and using ultrasonic tanks for storage. The main store consists of 32 tanks, each of which is about 5 ft. long and holds 32 numbers of 17 binary digits, one being a sign digit. This gives 1024 storage locations in all. It is possible to run two adjacent storage locations together so as to accommodate a number with 35 binary digits (including a sign digit); thus at any time the store may contain a mixture of long and short numbers. Short tanks which can hold one number only are used for accumulator and multiplier registers in the arithmetical united, and for control purposes in various parts of the machine.

      A single address code is used in the EDSAC, orders being of the same length as short numbers.

SOURCE: "The Origins of Digital Computers" Edited by B. Randell. Springer-Verlag, Berling (1973) p. 389.

SEE ALSO: "The design of a practical high-speed computing machine" by M. V. Wilkes, Proc. Roy. Soc. London A195 (1948) p. 274.

ALSO: "Automatic digital calculators" by M. V. Wilkes. Methuen, London (1956)

1949   COLD CATHODE STEPPING TUBE   Remington Rand (U. S. A.)

      The first published account of a multi-cathode stepping tube described a tube developed in America by Remington Rand. The first tubes to be widely used, however, were made in England by S. T. C. and by Ericsson Telephones. For some ten years following its introduction in 1949, the Ericsson 'Dekatron' became practically synonymous with the cold cathode stepping tube. The original double-pulse 'Dekatron' was followed in 1952 by the single-pulse tube operating up to 20 kc/s. Routing guides were added in 1955 to simplify the construction of reversible scalers. In 1962 auxiliary-anode tubes were introduced from which numerical indicators could be driven directly.

SOURCE: "A Survey of Cold Cathode Discharge Tubes" by D. M. Neale
The Radio and Electronic Engineer. February 1964. p. 87.

SEE ALSO: "Poly-cathode glow tube for counters and calculators"
by J. J. Lamb and J. A. Brustman. Electronics 22. No. 11 p. 92-6.
November 1949.

1949   DIP SOLDERING OF PRINTED CIRCUITS   S. F. Danko and
                                                    Abramson   (U. S. A.)

        When Danko and Abramson of the Army Signal Corps invented dip
soldering in 1949, a new era of automation came into being.

SOURC E: "Packaging and Materials" by R. L. Goldberg. Electronic Design
24. (Nov. 23, 1972) p. 126.

SEE ALSO: "Autosembly of miniature military equipment" by S. F. Danko and
S. J. Lanzalotti. Electronics 24 (7) (July 1951) p. 94.

ALSO: "Printed circuits and microelectronics" by S. F. Danko   Proc. IRE
(May 1962) p. 937.

ALSO: Autosembly.   U. S. Patent No:2756485 assigned to U. S. Army. July
31, 1956.

1949/   ION IMPLANTATION IN SEMICONDUCTORS   R. S. Ohl   (U. S. A.)
  50   W. Shockley

        Ion implantation is a technique for modifying the properties of solids
by injecting (implanting) charged atoms (ions) into them. The ions alter the
electrical, optical, chemical, magnetic and mechanical properties of a solid by
the interactions they have with the solid both as they slow down and by their
presence after they have come to rest.
        The implantation idea isn't a new one - work at Bell Laboratories
during the late 1940s and early 1950s by Russel S. Ohl and William Shockley
pioneered the application of ion implantation to semiconductor device fabrication.

SOURCE: "Ion implantation" by W. C. Brown and A. U. MacRae. Bell Laboratories Record (Nov. 1975) p. 389.

SEE ALSO: "Forming semiconductive devices by ionic bombardment"
W. Shockley. U. S. Patent No:2787564 (28 Oct. 1954)

ALSO: "Ion implantation in semiconductor device technology" by J. Stephen.
The Radio Electronic Engineer. Vol. 42 No. 6. (June 1972) p. 265.

1950   COMPUTERS (IBM 650)   I. B. M. (U. S. A.)

        The IBM 650, an intermediate-size, vacuum-tube computer was con-
sidered a workhorse of the industry during the late 1950's. Development began
in 1949 and the first installation was made late in 1954. Over a thousand 650's
have been in service since then.

        The 650 operates serially by character on words of 10 decimal digits plus sign. A 2-out-of-5 decimal representation in storage is translated into a biquinary code in the operating registers, allowing a fixed-count check to detect the presence of more or less than 2 bits per character. The main store of the 650 is a 12,500 rpm 2000-word magnetic drum. A two-address instruction format accommodates, as part of each instruction, the location of the next programme step. This format allows the programmer to place instructions anywhere on the program drum, and makes it possible for him to minimise access times to successive instructions.

SOURCE: Serrell, Astrahan, Patterson and Pyne "The evolution of computing machines and systems" Proc. IRE May 1962 p. 1050.

SEE ALSO: "The computer from Pascal to von Neumann" by H. H. Goldstine. Princeton Univ. Press 1972. p. 330.

1950   COMPUTERS (IBM 701)   I. B. M. (U. S. A.)

The development of the IBM 701 Data Processing System began at the end of 1950. A model was operating late in 1951 and the first production machine was delivered at the end of 1952. The heart of the IBM 701 system was a 36-bit single, address, binary, parallel, synchronous processor employing vacuum-tube flip-flops and diode logic at a rate of one megapulse per second. Multiple pluggable circuit packages were used. The arithmetic registers employed a recirculating-pulse bit-storage circuit, developed for the NORC in which a conbination of diode gating and pulse delay made it possible to store, shift right, or shift left with one triode per bit. Computation was governed by a single-address stored program of two 18-bit instructions per 36-bit word.

SOURCE: Serrell, Astrahan, Patterson and Pyne. "The eveolution of computing machines and systems" Proc. IRE May 1962. p. 1050.

SEE ALSO: "The computer from Pascal to von Neumann" by H. H. Goldstine. Princeton Univ. Press 1972. p. 265

1950's   TRANSISTORS - (Thermo-   O. L. Anderson
         compression bonding)   H. Christensen   (U. S. A.)
                                         and P. Andreatch

In the 1950's, O. L. Anderson, Howard Christensen and Peter Andreatch of Bell Laboratories discovered a new bonding technique particularly useful for connecting transistors to other elements in electronic circuits. The technique, pressing the connecting wire to the transistor mounting at low heat levels, provides a firm bond without introducing undesired electrical properties and has been widely used throughout the electronics industry. It is particularly advantageous in avoiding contamination, thus achieving long life and reliability.

SOURCE: " Mission Communications -- the story of Bell Laboratories" by Prescott C. Mabon. Published by Bell Laboratories Inc., Murray Hill, New Jersey U. S. A. (1975) p. 173.

1950's <u>FERREED SWITCH</u>   Bell Laboratories (U.S.A.)

The ferreed switch, invented at Bell Laboratories in the late 1950's, comprises two or four sealed-in-glass contacts and is controlled by magnetized wire coils. Ferreed switches, used to switch phone calls in most electronic switching systems, are smaller, faster operating and require less power than older switching devices.

<u>SOURCE</u>: "Mission Communications - the story of Bell Laboratories" by Prescott C. Mabon. Published by Bell Laboratories Inc., Murray Hill, New Jersey U.S.A. (1975) p.177.

1951   <u>QUALITY CONTROL</u>   J. M. Juran (U.S.A.)

First edition of "Quality Control Handbook" written by J. M. Juran, published by McGraw-Hill, New York.

An authorative treatise on all aspects of quality control.

1951   <u>CIRCUITRY (Super regeneration)</u> E. H. Armstrong (U.S.A.)

Superregeneration was discovered by E. H. Armstrong during the defense of his patent case for the regenerative receiver. In a superregenerative receiver, sustained oscillations are squelched by periodic variation of the effective resistance of the input resonant circuit. Oscillations periodically build up in a circuit resonant at the signal frequency. Sustained oscillations are prevented by periodic application to the grid of the superregenerative tube of a signal that damps the oscillations. The quenching frequency is usually between 20,000 and 100,000 cps. The superregenerative detector, because of its broad tuning and considerable sensitivity, was practical and popular earlier when unstable modulated oscillators were used as transmitters at frequencies above 30 Mc.

<u>SOURCE</u>: "Radio receivers - past and present" by C. Buff. Proc. IRE (May 1962) p. 887.

<u>SEE ALSO</u>: "Super-regenerative receivers" by J. R. Whitehead. Cambridge Univ. Press (1950)

1951   <u>AUTOMATIC CIRCUIT ASSEMBLY</u>   Nat. Bureau of Standards (U.S.A.)
        <u>"TINKERTOY" system</u>.

In 1950 the Navy Bureau of Aeronautics asked the National Bureau of Standards to study further automation of circuit assembly. The process that followed in 1951 - developed by Robert Henry of the Bureau of Standards - was dubbed Project Tinkertoy. It provided for the automatic assembly and inspection of circuit components, and it led to the first modular package.

The system started with individual components mounted on steatite ceramic wafers 7/8-inch square by 1/16-inch thick. The components were machine-printed or mounted over printed wiring. Four to six wafers were then automatically selected, stacked and mechanically and electrically joined by machine-soldered riser wires, which were attached at notches along the sides of each wafer. The resulting module generally had a tube socket on the top wafer.

Though this modular approach to packaging was used for production items, it faded in the late 50s as the transistor began to replace the vacuum tube.

Pie section modular build-up using Tinkertoy modules.

(Courtesy - Sanders Associates, Inc.)

SOURCE: "Solid State Devices - Packaging and Materials" R.L.Goldberg. "Electronic Design" 24. November 23, 1972. p.126.

1952   MICROELECTRONICS   G.W.A.Dummer   (U.K.)
(Integrated Circuit Concept)

In a paper read at the IRE Symposium in Washington D.C. on May 5th 1952, entitled: "Electronic components in Great Britain" G.W.A.Dummer stated:-

"At this stage, I would like to take a peep into the future. With the advent of the transistor and the work in semiconductors generally, it seems now possible to envisage electronic equipment in a solid block with no connecting wires. The block may consist of layers of insulating, conducting, rectifying and amplifying materials, the electrical functions being connected directly by cutting out areas of the various layers".

SOURCE: Proc.IRE Symposium on "Progress in Quality Electronic Components" Washington D.C. May 1952. p.19.

SEE ALSO: "A History of Microelectronics Development at the Royal Radar Establishment" by G. W. A. Dummer. "Microelectronics and Reliability" Vol. 4 (1965) p. 193.

ALSO: "Solid Circuits" - ' Wireless World. November 1957.

ALSO: "The Semiconductor Story - 3, Solid Circuits - a New Concept". by K. J. Dean and S. White. Wireless World. March 1973. p. 137.

1952   TRANSISTORS - ZONE MELTING   W. G. Pfann (U. S. A.)
        of GERMANIUM & SILICON

W. G. Pfann discovered a simple method for repeating the action of normal melting and freezing, which avoided handling the material between each operation. This resulted in material of extremely high purity which was then grown into single crystals by the pulling technique. Pfann also developed the zone levelling technique, which distributes impurities uniformly through a rod. He grew single crystals in his zone levelling apparatus using seeding techniques. The combination of zone levelling and horizontal growth of single crystals has become the standard technique used in today's transistor manufacturing operations.

SOURCE: "Contributions of materials technology to semiconductor devices" by R. L. Petritz Proc. IRE (May 1962) p. 1027

SEE ALSO: "Segregation of two solutes, with particular reference to semiconductors" by W. G. Pfann. J. Metals Vol. 4. (Aug. 1952) p. 861.

ALSO: "Techniques of zone melting and crystal growing" Solid State Physics Academic Press, New York Vol. 4. (1957) p. 423.

1952   DARLINGTON PAIRS, DIRECT-CONNECTED   S. Darlington   (U. S. A.)
        TRANSISTOR CIRCUIT

In one illustrative embodiment of this invention, a translating device comprises a pair of similar junction transistors the collector zones of which are electrically integral and the base zone of one of which is tied directly to the emitter zone of the other. Individual connections are provided to the other emitter and base zones. The device constitutes an equivalent single transistor having emitter and collector resistances substantially equal to those of one of the component transistors, but having a current multiplication factor substantially greater than that of either of the components.

SOURCE: "Semiconductor Signal translating device" U. S. Patent No: 2, 663, 806 Bell Telephone Laboratories dated May 9th, 1952.

1952   COMPUTERS (SAGE)   I. B. M.   M. I. T. (Lincoln Labs.)   (U. S. A.)

Descendents of M. I. T's "Whirlwind I' and of the IBM 701, the AN/FSQ -7 air defence computers for the SAGE system began, in 1952, as a co-operative IBM-M. I. T. Lincoln Laboratories effort based upon previous studies and specifications by Lincoln Laboratories. SAGE, a real-time communication-based digital computer control system, accepts radar data over phone lines, processes, displays information for operator decisions and guides interception weapons. The

first engineering model of the computer was delivered by IBM in 1955, and production deliveries began in June 1956.

SOURCE: Serrell, Astrahan, Patterson and Pyne "The evolution of computing
machines and systems" Proc. IRE May 1962. p. 1051.

1952   TRANSISTORS (single crystal
        fabrication) - Silicon.   G. K. Teal and E. Buehler   (U.S.A.)

        Large single crystals of silicon and silicon p-n junctions were pre-
pared by Teal and Buehler by an extension of the pulling technique developed
for germanium. These crystals were used by G. L. Pearson to prepare p-n
junction diodes by the alloy method.

SOURCE: "Contributions of materials technology to semiconductor devices"
by R. L. Petritz. Proc. IRE (May 1962) p. 1028.

SEE ALSO: "Growth of silicon single crystals and of single crystal p-n
junctions" by G. K. Teal and E. Buehler. Phys. Rev. Vol. 87. (July 1952) p. 190.

1952   DIGITAL VOLTMETER   A. Kay (U.S.A.)

        The digital revolution started in 1952, when Andy Kay unveiled the
first digital voltmeter. The Model 419 was crude, compared with today's DVMs.
But both the idea and Non Linear Systems - the company formed around the
idea - took off like a rocket. Today, almost every instrument from signal
generators to multimeters to scopes is digitized, thanks for Andy Kay and to
the commercial digital readout tube, introduced by Burroughs (then Haydu)
just one year before. (Burroughs' familiar Nixie tube actually had a rival in
its early days - the Inditron, which was developed by National Union Radio
Corp. and which did not survive.)

SOURCE: "Solid State Devices - Instruments" S. Runyon.
"Electronic Design" 24. November 23, 1972. p. 102.

1952   NEGATIVE - FEEDBACK TONE   P. J. Baxandall (U.K.)
        CONTROL CIRCUIT

        The circuit to be described is the outcome of a prolonged investigation
of tone-control circuits of the continuously-adjustable type, and provides independent control of bass and treble response by means of two potentiometers,
without the need for switches to change over from "lift" to "cut". Unusual
features are the wide range of control available, and the fact that a level response is obtained with both potentiometers at mid-setting. The treble-response
curves are of almost constant shape, being shifted along the frequency axis
when the control is operated, and there is practically no tendency for the curves
to "flatten off" towards the upper limit of the audio range. The shape of the
bass-response curves, though not constant, varies less than with most continuously-adjustable circuits.

SOURCE: "Negative-feedback tone control" by P. J. Baxandall. Wireless
World. (Oct. 1952) p. 402.

1953   <u>TRANSISTOR (Surface barrier)</u>   Philco (U.S.A.)

Advancing the early trend toward higher frequencies, Philco
developed the jet-etching technique in 1953. Here electrochemical machining
was used to fabricate the necessary thin base layers. A major product of
this process was the surface-barrier transistor, which boosted the upper
frequency limit of transistors into the megahertz region.

<u>SOURCE</u>: "Solid State Devices - The Processes" E.A.Torrero.
"Electronic Design" 24. November 23, 1972. p.73.

1953   <u>MASER (Microwave amplification by</u>   C.H.Townes and
        <u>stimulated emission of radiation)</u>   J.Weber   (U.S.A.)
                                                N.G.Basov and
                                                A.M.Prokhorov (U.S.S.R)

The first clear recognition of the possibility of amplification of
electromagnetic radiation by stimulated emission seems to have been by a
Russian, Fabrikant, who filed a patent in 1951 (although it was not published
until 1959) and who had discussed various aspects of his thesis of 1940.
However, his attempts to produce optical amplification in caesium were
unsuccessful.

The first statement in the open literature about amplification was
by Weber in 1953, followed by the detailed proposals of Basov and Prokhorov
f or a beam-type maser in 1954. However, the real excitement was caused
by the short article of Gordon, Zeiger and Townes, in the same year,
announcing the operation of the first maser using ammonia. Townes had
conceived the required experimental arrangement three years earlier, based
on his experience in microwave spectroscopy. In the years immediately
following many other techniques were studied, but the only one to give any
degree of practical success was the three-level maser of Bloembergen which
resulted in the ruby maser amplifier.

<u>SOURCE</u>: "Lasers and optical electronics" by W.A.Gambling. The Radio &
Electronic Engineer. Vol.45. No.10.(Oct.1975) p.538.

<u>SEE ALSO</u>: "Evolution of masers and lasers" by B.A.Lengyel.
V.A.Fabrikant. Am.J.Phys. 34. pp.903513. 1966.
<u>ALSO</u>: "Amplification of microwave radiation by substances not in thermal
equilibrium" by J.Weber. IRE Trans.on Electron Devices. No_ PGED-3,
p.1-4. June 1953.

<u>ALSO</u>: "Application of melecular beams to radio spectroscopic studies of
rotation spectra of molecules" by N.G.Basov and A.M.Prokhorov. J.Exp.
Theor.Phys.(U.S.S.R) 27, pp.431-8. 1954.

<u>ALSO</u>: "Molecular microwave oscillator and new hyperfine structure in the
microwave spectrum of NH_3" by J.P.Gordon, H.J.Zeiger and C.H.Townes.
Phys.Rev.95. pp.282-4. 1st July 1954.

ALSO: "Proposals for a new type solid-state maser" by N. Bloembergen
Phys. Rev. 104, pp. 324-7, 15th October, 1956.

1953   TRANSISTORS (Floating zone refining   P. H. Keck, R. Emeis &
        of silicon)   (U.S.A.)
                                                      H. C. Theurer

        An improved silicon purification technique was developed which pro-
duced material of sufficient quality that alloy silicon transistors could be fabricated with good yields. This was a novel variation of zone refining called
"floating zone refining" developed by P. H. Keck and independently by R. Emeis
and H. C. Theurer. This operation employs a vertical system and uses surface
tension to support a stable liquid zone formed by induction heating. Hence, the
crucible is completely eliminated. Silicon with thousands of ohm-cm resistivity
and minority carrier lifetimes of greater than 100 μsec can be produced by this
method.

SOURCE: "Contributions of materials technology to semiconductor devices"
by R. L. Petritz. Proc. IRE (May 1962) p. 1028.

SEE ALSO: "Crystallisation of silicon from a floating liquid zone" by P. H. Keck
and M. J. E. Golay. Phys. Rev. Vol. 89 (March 1953) p. 1297.

ALSO: "Growing single crystals without a crucible" Z. Naturforsch. Vol. 9A
(January 1954) p. 67.

ALSO: "Removal of boron from silicon by hydrogen water vapor treatment"
J. Metals Vol. 8. (October 1956) p. 1316.

1953   CONNECTION TECHNIQUES   R. F. Mallina et al (U.S.A.)
        (Wire wrapped joints)

        The mechanical basis of the wire wrapped joint was investigated very
extensively by workers at Bell Telephone Laboratories 20 years ago. Their
very full analysis of the joining system is still considered to be essentially
correct. The work includes photoelastic observations on a wrapped joint model
to investigate strain patterns produced by wrapping, and the study of stress
relaxation as a function of time and temperature.

        The wire wrapped joint consists of a wire which is tightly wrapped
around a sharp cornered terminal. Sufficient deformation is engendered in the
many notches created by the terminal in the wrapping wire to create metal to
metal interfaces with a high level of integrity. The wrapping wire, which is
bent several times during wrapping before final positioning in the wrap, is under
a high level of tensile stress during wrapping. The tensile strain which is
caused remains in the wire after wrapping because the stretched wire is locked
by the notches formed in it.

SOURCE: "Wire wrapped joints - a review" by P. M. A. Sollars. Electro-
component Science and Technology. 1974. Vol. 1. p. 17.

SEE ALSO: "Solderless wrapped connections" The Bell System Technical
Journal (May 1953)   Introduction: J. W. McRae
                        Part 1   - Structure & Tools - R. F. Mallina.

Part 2   - Necessary conditions for obtaining a
permanent connection - W. P. Mason and
T. F. Osmer.

Part 3   - Evaluation & performance tests - R. H. Van
Horn.

1953   <u>COMPUTERS IBM 704, 709 and 7090.</u>   I. B. M.   (U. S. A.)

The development of the IBM 704, descended from the 701, began in
November 1953 and the first system was delivered in January 1956. The 704
featured higher speed, magnetic-core memory, floating point and indexing. It
was followed in 1958 by the 709, featuring simultaneous read, write and compute
by means of Data Synchronizer units that allowed the input-output channels to
operate independently, as well as several special operations, including a table
look-up instruction and indirect addressing. A 32, 768-word core memory was
installed on a 704 in April, 1957.

The IBM 7090, the first units of which were delivered in 1959, is a
transistorised system compatible with the 709. About 5-times faster than the
709 on typical problems, the 7090 incorporates a $2.18-\mu sec$ core memory and
improved magnetic-tape units.

<u>SOURCE:</u>   Serrell, Astrahan, Patterson and Pyne. "The evolution of computing
machines and systems" Proc. IRE May 1962. p. 1052.

<u>SEE ALSO</u>:   "The computer from Pascal to von Neumann" by H. H. Goldstine.
Princeton Univ. Press 1972.

1953   <u>AUTOMATIC ASSEMBLY SYSTEMS</u> Autofab (General Mills)   (U. S. A.)

IBM Machine (IBM)   (U. S. A.)

United Shoe Machinery Co.(U. S. A.)

GE/Signal Corps.   (U. S. A.)

Mini-Mech (Melpar)   (U. S. A.)

In attempting to develop standards which would allow unrestricted
electronics design and production processes sufficiently flexible to permit
rapid transition from design to manufacturing, the following conclusions are
reached: -

1.   The use of a printed-circuit board is basic to practically all
automation approaches.

2.   A generic grid pattern to govern circuit layout must be used in
order to prevent obsolescence of tooling (0.1 and 0.025 in. are
commonly accepted.)

3.   In conjunction with the standard grid pattern, standard mounting
dimensions for all components and parts must be used.

The use of single-head component-insertion machines will satisfy these
fundamental requirements. The flexibility of such a machine permits the
insertion of a variety of components by recycling the printed-circuit modules
through the unit for each component. When higher volumes are obtained,
several machines can be utilized. By this evolutionary method one is able to

create any degree of automation which the manufacturer wishes to attain. It is
possible to stop with the production of the printed circuit and to use one or
several component-attaching or inserting machines. Dip-soldering can be
mechanised or performed manually. It would be impractical to consider taking
the giant step into completely automatic production at the beginning of the
factory's use of automation principles. Even for mass producers, it is
necessary to develop specific procedures for particular requirements documented by many years of experience in the step-by-step evolution toward a
mechanised operation.

SOURCE: Electronic equipment design and construction"by G. W. A. Dummer,
C. Brunetti and L. K. Lee. McGraw-Hill, New York (1961) p.185-186

SEE ALSO: Proceedings of Symposium on Automatic Production of Electronic
Equipment, sponsored by Stanford Research Institute and the U. S. Air Force
(April, 1954.)

General Mills "Autofab" in-line machine.

1953   TRANSISTOR (Unijunction)   UJT   G. E. C.   (U. S. A.)

        Engineers at General Electric's electronics advanced semi-
conductor laboratory in Syracuse, N. Y.,   experimented in the early 1950's
with germanium alloy tetrode devices in search of a semiconductor that could
be used at frequencies higher than the 5-megahertz operating level of
existing conventional bipolar transistors. The tetrode structures, it was
found, produced a transverse electric field that boosted the device's cutoff
frequency and lessened the semiconductor's input impedance - the limiting
factor regarding frequency.

        But in 1953, while examining the waveforms on the structure's
terminals I. A. Lesk of the laboratory noticed that an oscillatory signal was
present on the tetrode's emitter. And when the collector supply was removed,
the oscillations persisted for a while. The researchers realized that they had
stumbled on a new switching-type device.

        The GE engineers added to the development tetrode models, a
double-based diode structure that was being studied because of a negative-
resistance property.

SOURCE: "Solid state - a switch in time" by W. R. Spofford Jr., and
    R. A. Stasior. Electronics. February 19, 1968. p. 118.

1954   TRANSISTOR RADIO SET   Regency (U. S. A.)

        In 1954 the first transistor radio, the Regency, appeared on the
market. Although not a commercial success, it introduced the transistor
into the consumer market and gave transistor makers the impetus they
needed to develop mass production techniques. That, coupled with an
awakening of interest by the military, increased transistor sales meteor-
ically in the mid-fifties.

SOURCE: "Silicon, Germanium & Silver - the transistor's 25th anniversary"
by C. P. Kocher. The Electronic Engineer. November 1972. p. 30.

1954   SOLAR BATTERY   D. M. Chapin, C. S. Fuller
                                and G. L. Pearson.   (U. S. A.)

        As an outgrowth of work on transistors, Bell Laboratories scientists
D. M. Chapin, C. S. Fuller and G. L. Pearson in 1954 invented the silicon solar
battery - an efficient device for converting sunlight directly into electricity.
Arrays of these devices are used to power satellites and as energy sources
for other uses.

SOURCE: "Mission Communications -- the story of Bell Laboratories" by
Prescott C. Mabon. Published by Bell Laboratories Inc., Murray Hill,
New Jersey, U. S. A. (1975) p. 172.

1954   <u>TRANSISTOR (Interdigitated)</u> N.H.Fletcher (U.S.A.)

      The father of the interdigitated transistor is N.H.Fletcher, an engineer with Transistor Products Inc. When, in 1954, he hit upon the idea of elongated emitter areas, Fletcher was seeking a means to increase the power handling capability of devices, not a way to boost their cut-off frequency levels.

      His discoveries were applied by other firms to most transistor types over the next decade, but his own company realised few benefits from his work.

<u>SOURCE</u>:   "Solid state - fingers in the die" by J.E.Tatum. Electronics. February 19,1968. P.94.

<u>SEE ALSO</u>:   "Some aspects of the design of power transistors" by N.H. Fletcher. Proc. IRE. May 1955. p.551.

1955   <u>CRYOTRON</u>   D.Buck (U.S.A.)

      A super-conducting switching element was first examined by de Haas and Casimir-Jonker in 1935; however, it was not until 1955 that Buck demonstrated a practical device which he called the cryotron. The basic principle of the cryotron depends on the existence of a critical magnetic field above which the superconducting metal becomes a normal conductor. The original cryotron utilized a small tantalum rod wound with a niobium wire. At liquid helium temperatures the tantalum wire has a critical field of the order of several hundred gauss, whereas that of niobium is of the order of 2000 gauss. Consequently, when the niobium wire is pulsed with a suitable current, the magnetic field that it creates is sufficient to destroy the superconductivity in the tantalum but not in itself. The current in the niobium wire can be smaller than that in the tantalum wire so that a small current can control a larger one, thus producing a current gain in the device.

<u>SOURCE</u>:   "Solid State Devices'other than Semiconductors" by B.Lax and J.G.Mavroides. Proc.IRE (May 1962) p.1016.

<u>SEE ALSO</u>:   D.Buck "The cryotron - a superconductive computer component" Proc.IRE Vol.44. (April 1956) p.482.

1955   <u>INFRA-RED EMISSION FROM</u>   R.Braunstein   (U.S.A.)
    <u>GALLIUM ARSENIDE SEMI-</u>
    <u>CONDUCTORS</u>

      Radiation produced by carrier injection has been observed from GaSb, GaAs, InP, and the Ge-Si alloys at room temperature and $77^{\circ}K$. The spectral distributions of the radiation are maximum at energies close to the best estimates of the band gaps of these materials; consequently, the evidence is that the radiation is due to the direct recombination of electron-hole pairs.

<u>SOURCE</u>:   "Radiative transitions in semiconductors" by R.Braunstein. Phys. Rev. Vol.99. (1955) p. 1892.

1956   <u>TRANSISTORS (Diffusion process)</u>   C.S.Fuller & H.Reis (U.S.A.)

The next major advance in device technology was the diffusion process. Research on the diffusion of III-V impurities into germanium and silicon by Fuller at the BTL, and by Dunlap at GE laid the foundation for transistor fabrication using diffusion as a key process step. The BTL was the first to fully integrate these results into germanium and silicon transistors.   ;

Diffusion techniques have proved to be one of the best controlled methods for preparing p-n junctions. Because the common doping impurities diffuse very slowly in semiconductors at rates which can be varied by adjusting temperatures, close control and reproducibility of the impurity distributions can be achieved. Hence, control over the electrical parameters of the resulting devices may be maintained. The ability to form base regions only a fraction of a micron thick allows very high-frequency transistors to be fabricated.

<u>SOURCE</u>: "Contributions of materials technology to semiconductor devices" by R.L.Petritz. Proc.IRE (May 1962) p.1029.

<u>SEE ALSO</u>: "Diffusion processes in germanium and silicon" by H.Reis and C.S.Fuller. Chap.6 of "Control of composition in semiconductors by freezing methods" Ed.N.B.Hannay. Reinhold Pub.Corp.New York, 1959.

1956   <u>RADIO PAGING</u> Multitone (U.K.)

Concurrently with the development of two-way radio communication, there has been a remarkable development of radio-paging equipment and services. The first paging systems were established in the mid-fifties using a magnetic loop around the building to be served and operated on very low frequencies around 70 kHz. One of the first of these systems was installed by the Multitone Company at St.Thomas' Hospital, London in 1956.

Later the technique changed to v.h.f. radiating system using frequencies in the 27, 150 and 450 MHz bands. Development is now very widespread with over 2000 systems and 100,000 paging receivers in use in Britain alone. The receivers involved are very small, weighing only a few ounces.

<u>SOURCE</u>: "Fifty years of mobile radio" by J.R.Brinkley. The Radio and Electronic Engineer. Vol.45. No.10. (Oct.1975) p.556.

1956   <u>"FLOWSOLDERING" OF PRINTED</u>
         <u>CIRCUITS.</u>   Fry's Metal Foundries Ltd.   (U.K.)

In the Flowsolder dipping unit, developed by Fry's Metal Foundries Ltd, which avoids some of the difficulties inherent in the conventional flat dip-soldering of printed circuits, a stationary wave of molten solder is created by pumping the metal upwards through a rectangular nozzle and the pre-fluxed circuit panels are passed through the crest of the wave. It is claimed that this unit, which is used with specially developed fluxes, facilitates the soldering of printed circuits, free from faulty joints or bridging.

<u>SOURCE</u>: "Flowsolder Method of Soldering Printed Circuits" R.Strauss & A.F.C.Barnes. Electronic Engineering. Vol.28, No.345 (Nov.1956) pp.494-496.

1956   VALVES - VAPOUR COOLING   C. Beutheret   (France)

The first vapour-cooled tubes were made by Beutheret who used an anode with teeth approximately 10 mm square tapering to 5 mm square over 20 mm protruding from the surface. The object of the teeth was to stabilise the anode temperature and prevent a sudden catastrophic increase in anode temperature known as calefaction.

SOURCE: "Electronics Engineer's Reference Book" Newnes-Butterworth London(1976) Chap. 7.   p. 7 - 47

SEE ALSO: "The Vaportron Technique" by C. Beautheret. Rev. Tech. Thomson-CSF Vol. 24. (1956).

1956   MAGNETIC MATERIAL -   F. Bertaut and
         YTTRIUM IRON GARNET   F. Forrat   (France)

The crystallographic structure of yttrium iron garnet (YIG) was discovered by Bertaut and Forrat in 1956 and very soon afterwards large bulk crystals (several centimetres in one dimension) began to be grown by the molten flux technique. Yttrium iron garnet was found to be cubic, to be essentially an insulator, to have a saturation magnetization of 1750 gauss, and, unlike the previously available ferrite materials, to have a ferrimagnetic resonance linewidth of the order of 1 oersted at 10 GHz.

SOURCE: "Epitaxial magnetic garnets" by J. H. Collins and A. B. Smith. The Radio and Electronic Engineer. Vol. 45. No. 12. (Dec. 1975) p. 707.

SEE ALSO: "Structure of ferrimagnetic ferrites of rare earths" by F. Bertaut and F. Forrat. C. R. Acad. Sci. Paris, 242, p. 382. 1956.

1956   SEMICONDUCTOR DIODE   L. J. Giacoletto and J. O'Connell   (U.S.A.)
         JUNCTION CAPACITOR

A semiconductor juncti on when biased in the reverse (non-conducting) direction is a capacitance which can be varied by the bias voltage. When biased in this direction the mobile charge carriers are moved away from the junction, leaving uncompensated fixed charges in a region near the junction. The width, and hence the electrical charge of this space-charge layer, depends on the applied voltage, thus giving rise to a junction transition capacitance.

SOURCE: "History, Present Status and Future Developments of Electronic Components" by P. S. Darnell, IRE Transactions on Component Parts. September, 1958. p. 128.

SEE ALSO: "A variable capacitor germanium junction diode for U. H. F." by L. J. Giacoletto and J. O'Connell. R. C. A. Review. Vol. 17. March 1956. p. 68.

1957   <u>PLATED-WIRE MEMORIES</u>   U.F.Gianole   (U.S.A.)

       The plated--wire memory uses the principle of the direction of mag-
netization in a material to store digital information. The original concept of
the wire memory was invented in 1957 by U.F.Gianola of Bell Laboratories.
Plated-wire memories require no standby power, are non-volatile, inexpensive
to manufacture and will work in a high electrical noise environment.

<u>SOURCE</u>: "Mission Communications - the story of Bell Laboratories" by
Prescott C.Mabon. Published by Bell Laboratories Inc., Murray Hill,
New Jersey U.S.A. (1975) p.179.

1957   <u>SCANNING ELECTRON MICROSCOPY</u>   C.W.Oatley et al (U.K.)

       In 1957 a team of scientists at Cambridge University made a break-
through in electron probe microanalysers which gave Britain a lead in this
field that has so far been maintained.

       Before the wholly-British development of scanning techniques, speci-
mens had to be moved under static probes and the element distribution plotted
laboriously, and slowly. Scanning made it possible to display the information
on a TV-type viewing system.

       The following year, Tube Investments Research Laboratories found
the value of scanning X-ray microanalysis, developed by the team at the
Cavendish Laboratory was so great, that they built an instrument of their own.

       Early in 1959, the Cambridge Instrument Co., entered into an agree-
ment with TI to manufacture such instruments. Production started later that
year and the first Microscan was completed for the UKAEA, Aldermaston, and
shown in the Cambridge Instrument Company's London office at the time of the
Physical Society Exhibition in January, 1960.

       During the same period the company's efforts to improve the resolution
of Microscan led to the merging of their work with that of the University's
Engineering Department where scanning electron microscopes were being
studied. Stereoscan was the result.

       This instrument had a field of focus some 300 times greater than any
previous microscope, optical or otherwise, and produced dramatic results of
both rough and delicate surfaces alike. So revolutionary were the photographs
taken on this instrument, that the company had to arrange a special demonstration before microscopists were convinced that they were true pictures of
the surface.

<u>SOURCE</u>: "From Microscan to Stereoscan ... Cambridge keeping Britain in
front" by P.Slater. Electronics Weekly (Jan.10,1968) p.16.

<u>SEE ALSO</u>: "First international conference on Electron and Ion Beam Science
and Technology" Edited by R.Bakish. John Wiley (1965).

<u>ALSO</u>: K.C.A.Smith and C.W.Oatley. Brit. J.App.Phys.(1955) Vol.6. p.391.

1957 <u>RESISTORS - (Nickel-chromium thin film)</u>   R. H. Alderton and F. Ashworth   (U.K.)

It is well known that nichrome is the material most used today in thin-film resistors. Alderton and Ashworth stressed the importance of the following parameters: the source temperature, the degree of vacuum maintaining in the system, and the temperature of the receiving surface during deposition from the vapour phase. They state that stable films can only be made if the substrate temperature is greater than 350°C and the vacuum in the system better than 10^{-4} torr. They also stated that the maximum surface resistivity that produced stable films was 300Ω/□. These results are still used today as a guide in the production of nichrome films. Alderton and Ashworth measured the resistivity as a function of thickness and obtained a temperature coefficient of resistance from 100-200ppm/°C.

<u>SOURCE</u>: "Resistive thin films and thin film resistors - History, Science and Technology" by J. A. Bennett. Electronic Components(Sept.1964) p. 748.

<u>SEE ALSO</u>: "Vacuum deposited films of a nickel-chromium alloy" by R. H. Alderton and F. Ashworth. Brit. Jour. Applied Physics. Vol. 8. (1957) p. 205.

1957 <u>TRANSISTORS (Oxide masking process)</u>   C. J. Frosch (U.S.A.)

Another important technological advance in this period was the development of oxide masking for silicon by C. J. Frosch of BTL. He observed that a thermally grown oxide on silicon impeded the diffusion of certain impurities, including boron and phosphorus. This technique, coupled with photographic masking against etching, provides a powerful tool for silicon processing.

<u>SOURCE</u>: "Contributions of materials technology to semiconductor devices" by R. L. Petritz. Proc. IRE (May 1962) p. 1030.

<u>SEE ALSO</u>: "Surface protection and selective masking during diffusion in silicon" by C. J. Frosch and L. Derick. J. Electrochem. Soc. Vol. 104 (September 1957) p. 547.

1958 <u>VIDEO TAPE RECORDER</u> (Ampex (U.S.A.)

The first battery of 'video' tape recorders, a system called Ampex, was installed in the largest American television studios early in 1958. This system used tape moving at a speed of 200 inches per second but only half an inch wide; the recording being done on three tracks, two for storing the video signals and one for sound. A special machine for the cutting and editing of the tape had to be devised as it could be edited visually like cine film. Today, the majority of television programmes are recorded on videotape before transmission, in black-and-white as well as in colour; cine-film material shot for television can also easily be transferred on to videotape. Whether live, film or video, the viewer cannot detect the difference; the quality is equally high.

An important development of video recording is the 'canned' television programme for homes and schools. There are three rival systems. Two work with 'cassettes', which are inserted into a special replay unit plugged into the television set; one system uses Ampex tape, the other 8-mm. film with two

parallel tracks for sound and vi sion, electronically recorded. The third system
a British-German venture, works with a fast-rotating (1,500 r.p.m.) disc and
a pickup, providing a monochrome or colour programme of up to 12 minutes.

SOURCE: "A History of Invention" by E. Larsen. J. M. Dent & Sons, London
and Roy Publishers, New York, 1971. p. 330.

1958   TUNNEL DIODE   L. Esaki   (Japan) now (U.S.)

      First in chronological sequence came the tunnel diode, first
described by Esaki in 1958. Again with hindsight, what a beautifully simple
idea - to form a p-n junction between two such highly-doped regions that, in
equilibrium, the continuity of the Fermi level across the junction would
result in an energy barrier to the flow of carriers in the "forward" direction.
The device thus presents a high impedance at low forward bias, progresses
through a region of negative impedance and then into a fairly normal "forward"
region of positive resistance.

SOURCE: "Semiconductor Devices - portrait of a technological explosion."
by I. M. Mackintosh. The Radio and Electronic Engineer. Vol. 45. No. 10.
(Oct. 1975) p. 517.

SEE ALSO: "New phenomenon in narrow germanium p-n junctions"
by L. Esaki. Phys. Rev. Vol. 109 (1958) p. 603.

1958   LASER (Light amplification by   A. L. Schalow & C. H. Townes (U.S.A.)
        stimulated emission of radiation).

      Because of the great interest aroused by masers it was not until
1957 that further serious attention was given to the idea of producing an optical
version of the maser.
      In their classic article of 1958 on the principles of laser action
Schawlow and Townes suggested potassium vapour as a possible medium and
much effort was devoted to it but with no success. The reasons for this
f ailure were rather puzzling, especially as other workers later found caesium
vapour to behave as predicted. Another medium under consideration was ruby
(Cr^{3+} in Al_2O_3) although an internal report at Bell Telephone Laboratories
concluded that the existing material was much too poor to give any hope of
success and the experts of the time expected that the first laser would be
based on a gas. Great was the surprise and general jubilation therefore when
Maiman, who had persevered with ruby, achieved laser action in 1960.
Maiman's own jubilation was short-lived as the manuscript which he prepared
announcing his remarkable result was rejected by Physical Review Letters
and an historic scoop of scientific journalism was ahieved by the journals
"Nature"and"British Communications and Electronics"which carried the first
announcement in the established scientific literature.
      Some months later the helium/neon laser was successfully
operated and there followed over the next few years a tremendous explosion of
publications on laser transitions in hundreds of different materials and on the
properties of laser devices.

SOURCE: "Lasers and optical electronics" by W. A. Gambling. The Radio and Electronic Engineer. Vol. 45. No. 10. (Oct. 1975) p. 539.

SEE ALSO: "Infrared and optical masers" by A. L. Schalow & C. H. Townes Phys. Rev. Vol. 112 (Dec. 15, 1958) p. 1940.

ALSO: "Infrared and optical masers" in Quantum Electronics. C. H. Townes Ed. Columbia Univ. Press. New York. 1960.

ALSO: "Stimulated optical production in ruby" by T. H. Maiman. Nature. Vol. 187 (6th August 1960) p. 493.

1958   AUTOMATIC CIRCUIT ASSEMBLY   U. S. Army Signal Corps.   (U. S. A.)
"Micro-module" system

     By 1957 the goal for packaging had shifted from automation to miniaturization. Working with the Army Signal Corps., RCA suggested an approach that was similar to Tinkertoy's but with smaller wafers. Using wafers 310 mils square, spaced 10 mils apart, RCA encapsulated the assembled module with an epoxy resin to increase mechanical strength and provide environmental protection.

     With RCA as the prime contractor for an $18-million contract, the Signal Corps promoted micromodule as a standard package. A Signal Corps team headed by Daniel Elders, Stan Danko and Weldon Lane established a continuing development program for micromodule.

     The micromodule approach combined high packaging density, machine assembly and modular design. It was the first attempt at functional modular replacement, where the entire module was treated as a single component. The program established a compact universal packaging system using standard-shaped parts. But just as micromodule was gaining popularity in the early 60s, the IC deflated its chances of achieving sufficient volume for a competitive price.

SOURCE: "Solid State Devices - Packaging and Materials" R. L. Goldberg "Electronic Design" 24. November 23, 1972. p. 126/7.

SEE ALSO: Signal Corps. Contract DA-36-039-SC-76968 RCA Camden. N. J.

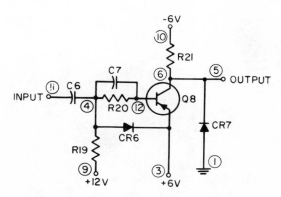

"Tinkertoy" original circuit.

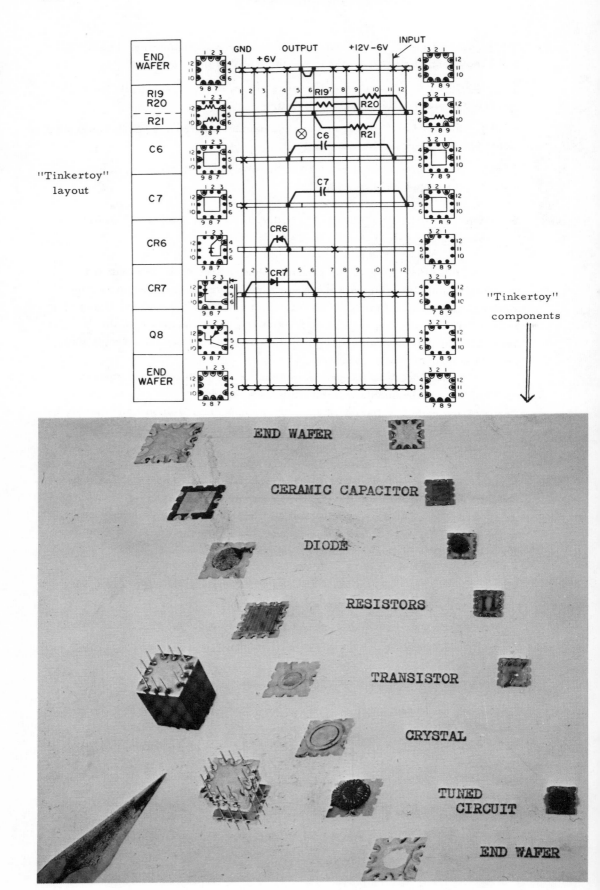

"Tinkertoy" layout

"Tinkertoy" components

1958   MOSSBAUER EFFECT   R. L. Mossbauer (Germany)

The Mossbauer effect is the phenomenon of recoiless resonance
fluorescence of gamma rays from nuclei bound in solids. The extreme sharp-
ness of the recoiless gamma transitions and the relative ease and accuracy in
observing small energy differences made the Mossbauer effect an important
tool in nuclear physics, solid state physics etc.

SOURCE: The Enclyclopaedia of Physics. (2nd Edition) Editor R. M. Besancon.
Von Nostrand. New York 1974. p. 595.

1958   TRANSISTOR (Field Effect (FET)   Stanislas Teszner. (France)

The first commercial FET was produced in France in 1958, by
Stanislas Teszner, a Polish scientist employed by CFTH, a General Electric
Company affiliate. Called the Tecnetron, Teszner's device was a germanium
alloy semiconductor. It had a transconductance of 80 micromhos, a pinchoff
of 35 volts, a gate leakage current of 4 microamps, and a low gate capacitance
of 0.9 picofarads. The low trans-conductance and high leakage severely
limited its applications. But the high pinchoff voltage was closer to the
operating levels of some tubes, and its gate capacitance permitted it to be
operated at a few megahertz.

SOURCE: "Solid state - an old-timer comes of age" by J. M. Cohen.
Electronics. February 19, 1968. p. 123.

1958   FIELD-EFFECT VARISTOR   Bell Laboratories   (U.S.A.)

This device, closely related in principle to the field-effect
transistor, has a constant-current feature which makes it ideally suited for a
current regulator in circuits where either the load or supply voltage varies
over wide limits. It can also be used as a current limiter or pulse shaper.
Its ac impedance is very high, making it useful as a coupling choke or as an
ac switch. This device is based on the field effect principles developed by
Shockley, Dacey and Ross.

SOURCE: "History, Present Status and Future Development of Electronic
Components" by P. S. Darnell. IRE Transactions on Component Parts.
September, 1958. p. 128.

SEE ALSO: "A Field Effect Varistor" Bell Labs. Record. Vol. 36.
April, 1958. p. 150.

1959   MICROELECTRONICS   J. S. Kilby (USA)
(INTEGRATED CIRCUIT-PATENT)

"It is, therefore, a principal object of this invention to provide a novel
miniaturized electronic circuit fabricated from a body of semiconductor
material containing a diffused p-n junction wherein all components of the
electronic circuit are completely integrated into the body of semiconductor
material."

SOURCE: U.S. Patent no:3, 138, 743 filed February 6th, 1959. (J.S. Kilby)

SEE ALSO: U.S. Patent No:3, 261, 081 patented July 19th, 1966. (J.S. Kilby and
Texas Instruments).

1959   MICROELECTRONICS - THIN FILMS   Bell Laboratories   (U.S. A.)
        (Tantalum)

            Tantalum film circuitry is a single material technology in that
capacitors, resistors, and elementary interconnections are all derived
from tantalum. Use of tantalum for this purpose is based on its chemical
and structural stability, and on its capability of being anodized to form
dielectrics for capacitors, and to protect and adjust resistors.

            In addition to the general value of tantalum film circuitry,
tantalum film resistors and resistance networks, especially when sputtered
in nitrogen, have independent interest as exceptional circuit elements.

SOURCE: "Developments in tantalum nitride resistors" by D.A. McLean
N. Schwartz and E.D. Tidd.   IEEE International Convention, March 26, 1964.

SEE ALSO: "Microcircuitry with Refractory Metals" by D.A. McLean
IRE Wescon Convention Record. Vol. 3. Part 6, pp. 87-91. 1959.

1959   INTRINSIC 10µ PHOTOCONDUCTORS   W.D. Lawson
        Cadmium telluride and Mercury   S. Nielsen   (U.K.)
        Telluride.   E.H. Putley
                                           A.S. Young

            The elements mercury, cadmium and tellurium have been purified, and
crystals of the compounds CdTe and HgTe, and of the mixed compounds CdTe-
HgTe have been prepared. X-ray and cooling-curve data have established that
CdTe and HgTe mix in all proportions to give single-phase crystals. Hall-
coefficient and conductivity measurements show that HgTe is a semi conductor
with a very low activation energy (~ 0.01 eV) and a high mobility ratio (~ 100).
HgTe is opaque to infrared radiation out to a wavelength of 38 µ, but the mixed
crystals show absorption edges which vary in position with composition from
$0.8µ$ in pure CdTe to 13 µ in crystals containing 90 per cent HgTe. Photoconductivity has been observed in filamentary detectors made from the mixed
crystals.

SOURCE: "Preparation and properties of HgTe and mixed crystals of
HgTe-CdTe" by W.D. Lawson, S. Nielsen, E.H. Putley and A.S. Young. J. Phys.
Chem. Solids. Vol. 9 (1959) p. 325.

1959   PLANAR PROCESS   J.A. Hoerni (U.S.A.)

            At Fairchild, Dr. Jean A. Hoerni, a physicist, was trying to
develop a family of double-diffused silicon mesa transistors. But instead of
mounting the base layer on top of the collector, the traditional mesa approach,
Hoerni diffused it down into the collector and protected the base-collector
junction on the top surface with a layer of boron-and-phosphorous diffused
silicon oxide. This first planar transistor was less brittle than the mesa and

far more reliable - dust or other foreign matter could not contaminate the
protected p-n junction. In 1959 Fairchild started marketing planar transistors
and shortly thereafter applied the planar technique to the new integrated
circuits.

SOURCE: "Silicon, Germanium & Silver - the transistor's 25th anniversary"
by C.P. Kocher. The Electronic Engineer. November 1972. p. 30.

1960   PRINTED WIRING MULTILAYER BOARDS   Photocircuits Corp.   (U.S.A.)

Miniaturized replacement for back-panel wiring in computers may be
accomplished by printed circuit sandwiches produced in many layers and lami-
nated together under heat and pressure. The components are being made by
Photocircuits Corporation, 31 Sea Cliff Ave., Glen Cove, New York.

The manufacturer believes the new development will have uses in circuits having multiple crossovers and complicated interconnections among closely
spaced component leads.

Connections between different levels in a multilayered circuit are made
through use of Tuf-Plate plated-through-holes. A typical six-layered printed-
circuit sandwich measures only 0.026 in. in thickness compared with a thickness
of 0.062 in. for a conventional single circuit board.

SOURCE: "On the market - P.C. sandwich six-layered unit" Electronics.
(April 8, 1960) p. 90.

1960   TRANSISTORS (Epitaxial crystal growth)   H.H. Loor, H. Christensen (U.S.A.)
                                                     J.J. Kleimock & H.C. Theurer

Until 1960 the semiconductor industry followed a pattern of starting
with a crystal as pure as needed in the initial stage, and each step added impurities in a controlled manner. In June, 1960, the Bell Telephone Laboratories
announced a new method of fabricating transistors using epitaxial single crystals
grown from the gas phase with controlled impurity levels. This technique had
been studied at a number of laboratories, but it was not until the 1960 announcement that the potential was fully grasped by the semiconductor industry. Its
unique advantage is the ability to grow very thin regions of controlled purity.

SOURCE: "Contributions of materials technology to semiconductor devices"
by R.L. Petritx. Proc. IRE (May 1962) p. 1030.

SEE ALSO: "New advances in diffused devices" by H.H. Loor, H. Christensen,
J.J. Kleimock and H.C. Theurer. Presented at the IRE/AIEE Solid State Device
Research Corp. Pittsburgh Pa. (June 1960)

ALSO: "Epitaxy - a fresh approach to semiconductor circuit design" by
Materials Dept. Motorola S/C Products Division. International Electronics
(March 1964) p. 24.

1960　　LIGHT EMITTING DIODE (LED)　　J. W. Allen and　　(U.K.)
　　　　　　　　　　　　　　　　　　　　　　　　P. E. Gibbons

It has been known for some time that rectifying contacts to GaP emit
light when current is passed through them (Wolff et al, 1955). Experiments at
this laboratory and elsewhere suggest that the electronic transitions involved
in this electroluminescence are different for the two directions of current flow.
We consider alloyed or point-contact junctions on n-type GaP. Then the light
emitted with forward bias has a spectrum which is a comparatively narrow band,
the position of the band depending on the impurities present in the GaP. If the
junction is biased in the reverse direction the current flowing is small until a
certain voltage is reached. Beyond this voltage the current increases rapidly
and orange light is emitted which has a very broad spectrum extending from the
infra-red down to the absorption edge or beyond (Loebner and Poor, 1959). It
would seem that electroluminescence in the forward direction is due to radiative
recombination of injected carriers via impurity levels, while that in the reverse
direction is due to emission by 'hot' carriers produced by an avalanche breakdown (Chynoweth and McKay, 1956).

SOURCE: "Breakdown and light emission in gallium phosphide diodes" by
J. W. Allen and P. E. Gibbons. Journal of Electronics Vol. VII No. 6. (Dec. 1959)
p. 518.

1960　　COMPUTERS - UNIVAC Solid State 80/90　　I. B. M. (U.S.A.)

The UNIVAC Solid State 80/90 was designed as a medium-sized data-
processing system. The term "solid-state" refers to the use of "Ferractor"
magnetic amplifiers and transistors. The system consists of a central
processor, a read-punch unit, a 450 card-per-minute card reader and a 600-
lines-per-minute printer. The card equipment can be obtained for either the
80-column or the 90-column punched-card system. The first installation was
made in January 1960.

SOURCE: Serrell, Astrahan, Patterson and Pyne "The evoluti on of computing
machines and systems" Proc. IRE May 1962. p. 1053.

1960　　COMPUTERS (CD 1604)　　Control Data Corporation　(U.S.A.)

The CDC 1604 is a general-purpose data-processing system manufactured by the Control Data Corporation. The first installation was made in
January 1960.

The entire system includes some 100, 000 diodes and 25, 000 transistors.
The internal number system is the binary, with a word length of 48 bits. There
are 62 24-bit one-address instructions (2 per word) each including a six-bit
operation code, a three-bit index and 15 bits for the address. Indirect addressing is built-in and six index registers are provided. Arithmetic is performed
with fixed or floating point in the parallel synchronous mode concurrently with
other operations. Addition requires 4. 8 to 9. 6 μsec, multiplication 25. 2 to
63. 6 μsec. including storage access. 32, 768 words of magnetic-core storage
are provided. Input-output equipment includes paper-tape typewriter, punched
cards, magnetic tape (up to 24 units) and a 667-1000 lines-per-minute printer.

SOURCE: Serrell, Astrahan, Patterson and Pyne "The evolution of computing
machines and systems" Proc. IRE May 1962. p. 1054.

1960   COMPUTERS (HONEYWELL 800)   Honeywell   (U.S.A.)

       The Honeywell 800 is a general purpose data-processing system capable of running as many as 8 distinct programmes simultaneously without special instructions. The first installation was made in December 1960.

       The system includes 30,000 diodes and 6000 transistors, excluding peripheral equipment. The internal number structure is binary and binary-coded decimal with a word length of 48 bits, or 12 decimal digits. These 48 bits are assignable to numerical, alphanumerical or pure binary information. There are 59 basic instructions, each consisting of a twelve -bit operation core and 3 twelve-bit addresses. Eight index registers are available for each of the 8 programmes which can be run concurrently. Other special-purpose registers are available. Arithmetic is performed in a synchronous parallel-serial-parallel mode, concurrently with other operations. Addition requires 24 μsec, multiplication 162 μsec including storage access. Up to 32,000 words of magnetic-core storage can be used. Input-output equipment includes punched cards, paper tape and a 900 lines-per-minute printer. Up to 64 magnetic tape units can be connected to the system.

SOURCE: Serrell, Astrahan, Patterson and Pyne "The evolution of computing machines and systems" Proc. IRE May 1962 p. 1054.

1960   SUB-MILLIMETER PHOTO-   E.H. Putley   (U.K.)
          CONDUCTIVE DETECTOR (n-type InSb)

       Photoconductivity has been observed using a cryostat fitted with a light pipe so that a specimen[*] could be illuminated with radiation of wavelength between 0.1 and 4.0 mm. The sources of radiation were a mercury lamp and grating spectrometer covering the range 0.1 to 1.4 mm. and a Philips DX151 klystron and garmonic generator operating at 2mm and 4mm. The light-pipe was fitted with a black paper filter at room temperature and a black polythene filter in the helium to remove short-wave radiation.

       The sample dimensions were 0.5 cm x 0.5 cm x 1.0 cm and indium electrodes were applied to opposite 0.5 cm x 1.0 cm faces. The magnetic field was applied at right angles to the direction of current flow and of the incident radiation. The radiation was directed along the long direction of the sample. For the majority of these experiments the radiation was modulated at 800 c/s and detected using a tuned amplifier and phase sensitive detector.

       When the temperature was reduced to below 1.5°K and a magnetic induction of 6-8000 gauss applied, the sample resistance was about 10-30kΩ. The sample was able to detect the applied radiation, the minimum detectable energy per unit bandwidth being approximately 5×10^{-10}W at 0.5 mm, 5×10^{-11}W at 2mm and 10^{-10}W at 4 mm.

SOURCE: "Impurity Photoconductivity in n-type InSb" Proc. Phys. Soc. Vol. 76 Pt. 5 No:491 (1 Nov. 1960) p. 802.

    * n-type indium antimonide.

1960–
1964

CIRCUITRY (logic circuits)   Various   (U.S.A.)

       Much of the early activity was involved with digital logic families. Almost from the beginning, a host of semiconductor manufacturers were attempting to establish the dominance of one logic family over the other – or were second-sourcing the strong suit of a competitor.

       At the start resistor-transistor logic (RTL) seemed to be the way to go; Fairchild and Texas Instruments were strongly promoting it. Then diode-transistor logic (DTL) appeared in 1962 from the recently formed Signetics, and it took off. Transistor-transistor logic (TTL) emerged in Sylvania's Universal High Level Logic (SUHL) in 1963 and again, more permanently, in Texas Instruments' 5400 series in 1964.

SOURCE: "Integrated circuits" E.A. Torrero – "Electronic Design" 24. November 23, 1972. p. 76.

1960

CIRCUITRY (Linear integrated circuits)   Various   (U.S.A.)

       Linear ICs came into their own in the 1960s. Starting with op amps, linear monolithics grew steadily in complexity and functions.

       Monolithic op amps were first introduced in the early 1960s. At least two manufacturers – Texas Instruments and Westinghouse – were selling models. Then Fairchild, in 1964, came out with the 702, the result of the first collaboration between Bob Widlar and Dave Talbert. The 702 found limited acceptance - more significantly, its development led to the 709, one of the biggest success stories in an industry accustomed to them.

       The 709 was a revolution of sorts. Rather than translate a discrete design into a monolithic form, the standard approach, Widlar played the linear microcircuit game by a different set of rules: Use transistors and diodes – even matched transistors and diodes – with impunity, but use resistors and capacitors – particularly those of large value – only where necessary. Even where use of a big resistor seemed inevitable, Widlar put a dc-biased transistor in its place. He exploited the monolith's natural ability to produce matched resistors and only assumed loose absolute values.

SOURCE: "Integrated circuits" E.A. Torrero – "Electronic Design" 24. November 23, 1972. p. 77.

1961

ELECTRONIC CLOCK   P. Vogel & Cie   (Switzerland)

       According to the present invention there is provided an electronic clock comprising no macroscopic moving parts which comprises an oscillator for delivering electrical pulses at a given frequency, distributing means arranged to be controlled by said oscillator for delivering at outputs thereof the pulses delivered by the oscillator, a counting device arranged to be controlled by the oscillator for delivering signals of a frequency of n cycles per hours, where n is an integral factor of 60, and 1 cycle per minute, an electronic switch arran-

ged to be controlled by the distributing means for delivering signals corresponding to the state of the counting device said signals being associated with
hours and minutes successively, to a distribution matrix for controlling a
display device.

SOURCE: British Patent Specification No:995, 546 "Improvements in or
relating to electronic clocks" Application made in the U.S.A. (No:94832) on
March 10th, 1961.

1961 TRANSFERRED ELECTRON EFFECT B.K.Ridley
                                    T.B.Watkins   (U.K.)

The possibility of obtaining negative resistance effects in a new way
in semiconductors is discussed. The principle of the method is to heat carriers in a high mobility sub-band with an electric field so that they transfer
when they have a high enough "temperature" to a higher energy low mobility
sub-band.

SOURCE: "The possibility of negative resistance effects in semiconductors"
by B.K.Ridley and T.B.Watkins. Proc.Physical Soc. Vol.LXXVIII (1961)p.293.

SEE ALSO: H.Kromer. Phys.Rev. Vol.109 (1958) p.1856.

1961 TRANSFERRED ELECTRON DEVICE   C.Hilsum (U.K.)

In some semiconductors the conduction band system has two minima
separated by only a small energy, and the lower minimum has associated with
it a smaller electron effective mass than the upper minimum. At high electric
fields it should be possible to transfer electrons to the upper minimum where
they will have a lower mobility. The conductivity of a homogeneous crystal bar
can therefore decrease as the field is increased, and it is conceivable that a
differential negative resistance could be obtained. The conditions needed for
obtaining negative resistance are examined, and calculations made for GaSb
and semi-insulating GaAs. It appears that negative resistances should be observable in both these materials.

SOURCE: "Transferred electron amplifiers and oscillators" by C.Hilsum.
Proc. of the IRE Vol.50. No.2 (February 1962). p.185.

SEE ALSO: "Proposed negative mass microwave amplifier" by H.Kromer.
Phys.Rev. Vol.109 (March 1958) p.1856.

1961 PRINTED WIRING (Multilayer boards
      with seeded plated through holes)   Hazeltine Corp. (U.S.A.)

A multiplaner printed circuit board consisting of a composite board
comprising a stack of insulative sheets bonded together, the sheets having
desired conductors configurations adhered thereto on surfaces which are
interior in the composite board, the board having continuous plated-through
holes which electrically interconnect the conductor configurations existing at
the various planes of the composite board, said holes having been formed after
the bonding together of the insulative sheets and then plated through by a process of seeding and then plating conductor material on the wall surfaces defin-
ing the holes.

SOURCE: U.S.A. Patent No 911, 718 dated 27th April, 1961.

1961   LIQUID PHASE EPITAXY   H. Nelson   (U.S.A.)

    An apparatus and procedures have been developed for the epitaxial growth of GaAs and Ge from the liquid state. The resulting technology has been found to possess advantages over vapour-phase epitaxy in some applications demanding highly doped epitaxial films and high-quality p-n junctions at the sub-strate-film interface. In this connection, it is an important feature of the liquid phase process that chemical impurities and mechanical damage of the substrate are removed when material is initially dissolved from the substrate surface prior to epitaxial growth. A clean interface p-n junction is thus obtained. Since liquid-phase epitaxy also favours the achievement of high doping and a steep concentration gradient at the p-n junctions, the process has proved itself eminently suitable for application in the manufacture of Ge tunnel diodes. In its application to the fabrication of the GaAs laser diode, it is an additional advantage of the liquid -phase process that the interface p-n junction is formed on a (100) crystal plane. As a consequence, this p-n junction is perfectly planar and also perpendicular to the (110) cleavage planes of the wafer. An optimum geometry (plane-parallel ends perpendicular to a perfectly flat p-n junction) is thus insured for diodes cleaved from (100) oriented GaAs wafers whose p-n junction has been formed by liquid-phase epitaxy.

SOURCE: "Epitaxial growth from the liquid state and its application to the fabrication of tunnel and laser diodes." by H. Nelson. RCA Review (Dec.1963) p603.

SEE ALSO: "Epitaxial growth from the liquid phase" by H. Nelson. Solid State Device Conference, Stanford University (June 26, 1961).

ALSO: "Properties and applications III-V compound films deposited by liquid phase epitaxy" by H. Kressel and H. Nelson. Physics of Thin Films. Vol. 7. Academic Press N. Y. (1973).

1962   COMMUNICATION SATELLITE   Various (U.S.A.)
        (Telstar I)

    The first earth satellite was launched by the USSR on the 4th October, 1957. Telstar I, the first communication satellite, successfully transmitted high-definition television pictures across the Atlantic on the 10th July 1962, and its successors promise a new form of global communication by sound and vision. Telstar I, now silent, has orbited the Earth about 17000 times and is expected to remain in orbit for some 200 years; the 170lb satellite was powered by nickel-cadmium batteries, recharged by 3600 solar cells, and contained 1064 transistors and a single electron tube (a travelling-wave tube for amplifying signals). By the 23rd July, 1962, 16 European countries were exchanging live television with the United States and Telstar 2, launched on the 7th May 1963, paved the way for the world's first commercial communication satellite, Early Bird.

SOURCE: "The Scope of Modern Electronics" by F. A. Benson. Electronics & Power (Jan. 1969) p. 13.

1962   <u>SEMICONDUCTOR LASER</u>   R. N. Hall   J. D. Kingsley
                                      G. E. Fenner   T. J. Soltys   (U.S.A.)
                                         and R. O. Carlson

       Coherent infrared radiation has been observed from forward biased
GaAs p-n junctions. Evidence for this behaviour is based upon the sharply
beamed radiation pattern of the emitted light, upon the observation of a threshold current beyond which the intensity of the beam increases abruptly, and upon
the pronounced narrowing of the spectral distribution of this beam beyond threshold. The stimulated emission is believed to occur as the result of transitions
between states of equal wave number in the conduction and valence bands.

<u>SOURCE</u>: "Coherent light emission from GaAs junctions" by R. N. Hall,
G. E. Fenner, J. D. Kingsley, T. J. Soltys and R. O. Carlson. Phys. Rev. Letters.
Vol. 9. No:9 (Nov. 1, 1962) p. 366.

1962   <u>SEMICONDUCTOR LASER</u>   M. I. Nathan
                                        G. Lasher   (U.S.A.)

       A characteristic effect of stimulated emission of radiation in a
fluorescing material is the narrowing of the emission line as the excitation is
increased. We have observed such narrowing of an emission line from a for-
ward-biased GaAs p-n junction. As the injection current is increased, the
emission line at $77^{\circ}K$ narrows by a factor of more than 20 to a width of less than
$kT/5$. We believe that this narrowing is direct evidence for the occurrence of
stimulated emission.

<u>SOURCE</u>: "Stimulated emission of radiation from GaAs p-n junctions" by
M. I. Nathan, W. P. Dumke, G. Burns, F. H. Dill Jr., and G. Lasher. App. Phys.
Letters. Vol. 1. No:1 (1. Nov. 1962) p. 62.

1962   <u>ELECTRONIC WATCH</u>   P. Vogel & Cie (Switzerland)

       An electronic timepiece, comprising an oscillator unit; a frequency
divider unit, and a time display unit, in which each unit comprises or consists
of a layer of semi-conductive material, the layers being sandwiched together
and having their interfaces insulated from each other except at selected points
at which the units are electrically connected together.

<u>SOURCE</u>: British Patent Specification No:1, 057, 453 "Electronic Timepieces"
Application made in Switzerland (No:13423) on Nov. 16th, 1962.

1962   <u>MICROELECTRONICS</u>   Y. Tao   (U.S.A.)
        (Flat-pack)

       With the emergence of the IC as the modern circuit element of
the early '60s, transistor packages were found to lack sufficient heat sinking
and adequate interconnections. To dissipate heat and provide a standard
package size, Yung Tao created the flatpack in 1962 while at Texas Instruments. It was 1/4 by 1/8 inch and originally had 10 leads.

<u>SOURCE</u>: "Solid State Devices - Packaging and Materials" R. L. Goldberg
"Electronic Design" 24, November 23, 1972. p. 127.

1962   <u>LED (LIGHT EMITTING DIODE)</u>   N. Holonyak (U. S. A.)

        Nick Holonyak, Jr. (F) a member of the faculty at the University of
Illinois at Urbana-Champaign, Ill. received the 1975 John Scott Award for his
inventions leading to the first practical light-emitting diode (LED).

        Dr. Holonyak received the bachelor's degree in 1950, the master's
degree in 1951 and the doctorate in 1954, all from the University of Illinois.
In 1962, while at the General Electric Advanced Semiconductor Laboratory,
Syracuse, N. Y., he invented the first gallium arsenide phosphide LED and
the first semiconductor laser to operate in the visible spectrum. Since
joining the University of Illinois faculty in 1963, he has continued his research
with semiconductors, diodes, and other solid-state devices.

<u>SOURCE:</u>   IEEE Spectrum.   October 1975. p. 106.

1963   <u>GUNN DIODE OSCILLATOR</u>   J. B. Gunn   (U. S. A.)

        The observation is described of a new phenomenon in the electrical
conductivity of certain III-V semiconductors. When the applied electric field
exceeds a critical value, oscillations of extremely high frequency appear in
the specimen current.

<u>SOURCE:</u> "Microwave oscillations of current in III-V semiconductors" by
J. B. Gunn. Solid State Communications Vol. 1 (1963) p. 88.

1963   <u>SILICON ON SAPPHIRE TECHNOLOGY</u>   Various (U. S. A.)

        The technology of silicon-on-insulating substrates, specifically
silicon on sapphire, dates back to the beginning of practical MOS technology
in 1963. The technology is known by different abbreviations, such as: SIS,
ESFI (epitaxial silicon films on insulators), SOS, SOSL (silicon on spinel), etc.

        The principal advantage of SOS circuitry is the inherent dielectric
isolation, both dc and ac. The absence of silicon, except in the active device
areas, significantly reduces parasitic capacitance between lines and essentially eliminates the parasitic capacitance to the substrate. Diffusion of device electrodes through the silicon film to the sapphire reduces electrode capacitance by several orders of magnitude because of the reduction in junction
area. This significant reduction in electrode and interelectrode capacitance
enables many devices to ahieve their maximum band-width and frequency response; it allows for very high speeds, minimum speed-power products (below
0. 5 pJ) on SOS CMOS and for very high frequency linear elements $(f_r > 2. 10^9$
Hz), such as dual-gate MOSFET's (tetrodes).

<u>SOURCE:</u> "Recent SOS technology advances and applications" by R. S. Ronen
and F. B. Micheletti. Solid State Technology(August 1975) p. 39.

<u>SEE ALSO:</u> Early publications on SOS Technology e. g. Material and Devices,
include: H. M. Manasevit and W. I. Simpson, "Single crystal silicon-on-Sapphire
substrate" J. Appl. Phys. 35, 1349 (1964) : C. W. Mueller and P. H. Robinson,
"Grown-film silicon transistors on sapphire" Proc. IEEE 52, 1487 (1964).

1964   <u>MICROELECTRONICS (DIP)</u>   B. Rogers (U.S.A.)
        <u>(Dual-in-line package)</u>

        Bryant (Buck) Rogers fostered the invention of the DIP while at
Fairchild Semiconductor in 1964. It originally had 14 leads and looked just
as it does today.

        <u>SOURCE:</u> "Solid State Devices - Packaging and Materials" R. L. Goldberg
"Electronic Design" 24, November 23, 1972. p. 127.

1964   <u>"IMPATT" DIODE</u>   R. L. Johnston and B. C. DeLoach (U.S.A.)

        In 1964, Bell Laboratories scientists R. L. Johnston and B. C. DeLoach
discovered the IMPATT (IMPact Avalanche Transit Time) diode, subsequently
shown to operate by an effect proposed earlier by W. T. Read Jr., also of the
Laboratories. IMPATT diodes - semiconductor devices that generate micro-
waves directly when a DC voltage is applied to them - are becoming increas-
ingly important in the design of microwave systems because of their high
reliability and low cost.

        <u>SOURCE:</u>   "Mission Communications - the story of Bell Laboratories" by
Prescott C. Mabon. Published by Bell Laboratories Inc., Murray Hill,
New Jersey, U.S.A. (1975) p. 173.

1964   <u>TRANSISTOR (Overlay)</u> R. C. A. (U.S.A.)

        The overlay transistors, first introduced in 1964, was developed at
RCA under a contract from the Army Electronics Command, Ft. Monmouth,
N. J., as a direct replacement for the vacuum tube output stages then used in
military transmitting equipment.

        The first commercial overlay, the 2N3375, produced 10 watts of
output power at 100 Mhz and could handle 4 watts at 400 Mhz. Comparable
interdigitated structures of that day were capable of 5 watts at 100 Mhz and
0.5 Mhz and 0.5 watts at 400 Mhz.

        <u>SOURCE:</u>   "Solid state - a worthy challenger for RF power honors" by
D. R. Carley. Electronics. February 19, 1968. p. 100.

        <u>SEE ALSO:</u>   "The overlay - a new UHF power transistor by D. R. Carley,
P. L. McGeough and J. F. O'Brian. Electronics August 23, 1965. p. 70.

1964   <u>MICROELECTRONICS (Beam Lead)</u>   M. Lepselter   (U.S.A.)

        In 1964, Martin Lepselter of Bell Telephone Laboratories
invented the beam lead as a mechanical and electrical interconnection
between the IC and its case.

        <u>SOURCE:</u>   "Solid State Devices - Packaging and Materials" R. L. Goldberg
"Electronic Design" 24, November 23, 1972. p. 127.

SEE ALSO: "Beam lead technology" by M. P. Lepselter. The Bell System
Tech. Journal. Vol. XLV No. 2. (Feb. 1966) p. 233.

1965   SELF-SCANNED INTEGRATED   G. P. Weckler   (U.S.A.)
PHOTODIODE ARRAYS

The possibility of forming image sensors from arrays of silicon photo-
diodes on a single silicon chip has been recognised for many years, probably
since the inception of integrated circuit technology some fifteen years ago. It
was quickly apparent that the array size was limited, not by the number of
diodes that could be included on the silicon, but by the number of output leads
necessary to form connections to these diodes. To circumvent this problem, it
was necessary to scan the diodes, that is, to multiplex them in to a single output lead by means of switching circuitry on the same integrated chip. A second
problem was that of detecting the minute photocurrents produced by the necess -
arily very small diodes. The technique of charge integration had been used in
the Vidicon for some years, and it was pointed out by G. P. Weckler in 1965 that
this technique could be used with photodiode arrays, the switching being achieved
by m.o.s. transistors. The one step necessary to complete the picture was now
to include a shift register on an integrated circuit with the diodes and m.o.s.t.s
to perform the serial multiplexing function. The first fully self-scanned arrays
using this technique were announced in 1967.

SOURCE: "Applications of self-scanned integrated photodiode arrays" by
P. W. Fry. The Radio and Electronic Engineer. Vol. 46. No. 4. pp. 151-160
(April 1976)

SEE ALSO: "Operation of p-n junction photodetectors in a photon flux integrat-
ing mode" by G. P. Weckler. IEEE J. Solid State Circuits, SC-2, No. 3. pp. 65-
73. (Sept. 1967)

ALSO: "Development and potential of optoelectronic techniques" by P. J. W.
Noble. Component Technology 2, No. 8. pp. 23-8, (December, 1967).

1965   MINICOMPUTER   Digital Equipment Inc. (U.S.A.)

It is generally accepted that the first minicomputer was designed by
Digital Equipment in 1965 - a 12-bit 4K word memory machine selling for approximately £15,000. Many applications were found for a machine of this type,
and the market blossomed rapidly with a number of manufacturers designing
specialised machines with 8, 12 or 16 bits and varying memory sizes . The
direct descendants of such machines, with increased power are still available
at prices around £1,500, i.e. a reduction of 10:1 over ten years. This price
reduction, and/or increase in performance, was made possible by the introduction of integrated circuits, MSI and SI logic, which allowed computers to
become physically smaller while at the same time increasing performance.
Since those early days when machines had very little software and peripheral
support great strides have been made with the addition of extras, such as disks
and magnetic tapes which allow the provision of operating systems running high
level languages.

SOURCE: "How minicomputers can produce an integrated solution to the
running of a business" by I. Evans. Paper read at Seminex, London (March 25,
1976)

1965   RADIO COMMUNICATION (Satellite)   Intelsat I

       The first internationally owned satellite, INTELSAT I, was put into operation in 1965. It was placed in a geo-stationary equatorial orbit, that is at an altitude of 22 400 miles, in a longitudinal position 30° West for trans-atlantic operation. It had a mass of 39 kg (85 lb) primary power 45 W from solar cells and was capable of relaying 120 voice circuits or one television channel. The INTELSAT I system was to some extent experimental for two main reasons. Firstly, it was to ascertain whether reliable communication could be maintained in spite of the high path loss of 200 dB; however, the Earth stations employed the now well-known parabolic reflector type aerials - diameter 85-100 ft - with cryogenic-cooled low-noise amplifiers and in this respect it was a great success. Secondly, it was to determine whether the transmission delay, Earth - satellite · Earth, of 250 ms was operationally acceptable.

       The decision was taken to continue with satellites in the geo-stationary orbit and this is used for all internationally-owned satellites today.

SOURCE: "Fixed Communications" by A.S. Pudner. Radio and Electronic Engineer. Vol. 45. No. 10. (Oct. 1975) p. 547.

1966   FLIP-FLOP BONDING TECHNIQUE   M. Wiessenstern   (U.S.A.)
                                            and G. A. S. Wingrove

       The flip-flop bonding structure and method was invented and subsequently patented in 1966 by Wiessenstern and Wingrove. Since that time nearly every semiconductor manufacturer has experimented with various forms of flip-flop bonding for the purpose of assembling integrated circuits, and possibly some discrete components, into larger subsystems. To this day, no successful method of flip-chip bonding has become generally utilized on the open market.

SOURCE: "A multichip package utilizing In-Cu flip-chip bonding" by A.P. Youmans, R.E. Rose and W.F. Greenman. Proc. IEEE. Vol. 57. No. 9. Sept. 1969. p. 1599.

SEE ALSO: "Semiconductor device assembly with true metallurgical bonds" by M. Weissenstern and G. A. S. Wingrove.   U.S. Patent No: 3256465 (June 14, 1966).

ALSO: "Joining semiconductor devices with ductile pads" by L. F. Miller. 3rd Ann. Hybrid Microelectronics Symposium (Oct. 29, 1968).

1966   COMMUNICATION (Optical Fibres)   K. C. Kao and G. A. Hockham (U.K.)

       Another method of providing guidance, and of cunningly circumventing the problem of light travelling in straight lines, is to use a fibre consisting of a glass core having a high refractive index surrounded by a cladding of lower index. As early as August 1964, in an address to the British Association for the Advancement of Science, the author speculated on the use of light and glass fibres in the telephone network, instead of electric currents and wires, but developments did not start in earnest until publication of the classic article of

Kao and Hockham of STL in 1966. At the time the problem seemed a formid-
able one; the attenuation of existing fibres was about 1000 dB/km, the bandwidth was expected to be low and fibre bundles were fragile. Since then
enormous strides have been made resulting in fibre attenuations of 2 dB/km
produced as a matter of routine, bandwidths of 1 GHz in a 1 km length of fibre
having a diameter of 100 μm, and fibres coated with nylon which are too strong
to be broken by hand. Such fibres are flexible and capable of being incorporated into simple but effective forms of cable. The bandwidth of a single fibre
is much greater, and the attenuation lower, than existing copper coaxial cables
and the diameter is considerably smaller. Thus the capacity of the present
telephone network could be very greatly increased, with little additional
installation expense, by the gradual introduction of optical fibre cables.

SOURCE: "Lasers and Optical Electronics" by W. A. Gambling. The Radio
and Electronic Engineer" Vol. 45. No. 10. (Oct. 1975) p. 541.

SEE ALSO: "Dielectric-fibre surface waveguides for optical frequencies"
by K. C. Kao and G. A. Hockham. Proc. Instn. Elect. Engrs. 113. pp. 1151-8.1966.

1967   LASER TRIMMING OF THICK FILM RESISTORS   --   (U.S.A.)

The trimming of electronic components started in 1967 when the first
experiments were conducted on the trimming of thick film resistors with CO_2
lasers. Two years later "Q-switched" laser systems were installed in the
General Motors Delco Electronics plant in Indiana for the manufacture of thick
film voltage regulators for automobiles.

SOURCE: "Bright future for laser trimming" by W. B. Cozzens.   Electronic
Engineering.   February 1976.   p. 58.

1967   SURFACE ACOUSTIC WAVE DEVICES   R. M. White (U.S.A.)

Lord Rayleigh first described the equations governing the propagation of
surface elastic plane waves along the stress-free boundary of a semi-infinite,
isotrophic and perfectly elastic solid. White's paper, in 1967, is summarised
as follows:

Surface elastic-wave propagation, transduction, and amplification (in a
piezoelectric semiconductor) are discussed with emphasis on characteristics
useful in electronic devices. Computed curves show the dependence on distance
from the surface of the elastic and the electric fields associated with surface
elastic-wave propagation in cadmium sulfide. The interaction impedance,
relating the external electric field to power flow, is computed for propagation
on the basal plane of CdS and found to be low in comparison with values characteristic of electromagnetic slow-wave circuits. Amplification with a continuous drift field in cadmium sulfide is reported, and differences between surface
and bulk-wave amplifiers are discussed. Some operating characteristics and
fabrication techniques for making electrode transducers on piezoelectric
crystals are given, together with experimental results on several passive
surface-wave devices.

SOURCE: "Surface elastic wave propagation and amplification" by R. M. White
IEEE Trans. on Electron Devices. ED-14 No. 4. (April 1967) p. 181.

SEE ALSO: "On waves propagated along the plane surface of an elastic solid" by Lord Rayleigh. Proc. London Math. Soc. Vol. 17 (Nov. 1885) p. 4.

ALSO: "Surface waves in anisotropic media" by J. L. Synge. Proc. Royal Irish Acad. (Dublin) Vol. A58 (Nov. 1956) p. 13.

ALSO: "Surface waves in anisotropic elastic media" by V. T. Buchwold. Nature. Vol. 191 (Aug. 26, 1961) p. 899.

ALSO: "Design of surface wave delay lines with interdigital transducers" by W. R. Smith, H. M. Gerard, J. H. Collins, T. M. Reader and H. J. Shaw. IEEE Trans. on Microwave Theory & Techniques. Vol. MTF17 No. 11 (Nov. 1969) p. 865.

ALSO: "Passive Interdigital Devices using surface acoustic waves" IEE Reprint Series 2. Ed. D. P. Morgan. Peter Peregrinus (May 1976).

1968   AMORPHOUS SEMICONDUCTOR SWITCHES   S. R. Ovshinsky (U. S. A.)

Switching phenomenon have been noted for a decade but Ovshinsky first attracted international attention by producing a discrete component commercially in 1968. Figure 7(a) shows an early switch made by his company in which a thin film of undisclosed glassy amorphous material is sandwiched between massive carbon electrodes in a thermistor type package. Another possible discrete geometry used in earlier studies is the cross-over sandwich using thin film electrodes (Fig. 7(b)).

These devices and their more modern counterparts are so-called threshold-switches, and their circuit configuration and resulting I-V characteristics are shown in Fig. 8. These devices have the remarkable property of having an initial OFF resistance, corresponding to region (a), of order tens of megohms, which drops to a much lower ON resistance, region (c) of around 100 Ω when the voltage across it exceeds a certain threshold value V_{th}, typically around 10 V. The switching, which occurs along a load line corresponding to the load resistor, region (b), happens in a time typically of order nanoseconds, but there is a delay time which can be as long as 10 μs, between the application of a switching voltage and the onset of switching. If the current in the On-state is allowed to fall below a minimum holding value, I_h, the device reverts to its high resistance state, along region (d). In this sense, the device behaves as a monostable element. Notice that the switch is insensitive to the polarity of the supply; the I-V curve for negative voltages is a mirror image of that for positive values.

SOURCE: "Amorphous semiconductor devices and components" by J. Allison and M. J. Thompson. The Radio & Electronic Engineer. Vol. 46. No. 1. (Jan. 1976) p. 12.

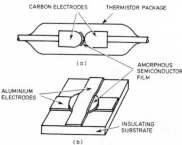

Fig. 7 Early discrete switches.

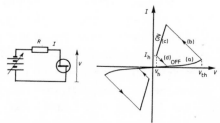

Fig. 8 Measuring circuit and voltage-current characteristic for a threshold device.

SEE ALSO: "Reversible electrical switching phenomena in disordered solids."
by S.R.Ovshinsky. Phys.Rev.Letts.Vol.21.(1968) p.1450.

1968   BARITT Diode   G.T.Wright   (U.K.)

In 1968 Wright described a new negative resistance microwave device
based on the principle of barrier controlled injection and transit time delay –
the BARITT diode. His simple analysis suggested that the device should operate at moderate power and low noise levels. In the same year, independently,
Ruegg presented a paper on the simplified large-signal theory of a similar
punch-through structure giving considerably optimistic prospects - an estimated
efficiency of the order of 20% and power output of 10-100 W at 10 GHz. These
theoretical works were confirmed experimentally in 1970 when Sultan and Wright
achieved negative resistance in npn silicon structures, and subsequently oscillations in pnp structures and in 1971 when Coleman and Sze reported oscillations
in metal-semiconductor-metal structures. Several experimental papers have
since been presented, comparing the properties of different BARITT diode
structures and pointing out the reliable and low-noise operation of the device at
moderate power levels.

SOURCE: "Large-signal analysis of the silicon pnp-BARITT diode" by
M.Karasek. Solid State Electronics Vol.19.(1976) p.625.

SEE ALSO: G.T.Wright Elect.Lett.4. 543 (1968)

ALSO: H.W.Ruegg, I.E.E.E.Trans.ED-15, 577 (1968)

1969   MAGNETIC BUBBLES   A.H.Bobeck, R.F.Fischer,
                              A.J.Perneski, J.P.Remeika   (U.S.A.)
                              and L.G.Van Uitert.

A magnetic material usually consists of arrays of discrete localized
volumes of material, defined as domains. Each domain separated from its
neighbours by domain walls has a preferential orientation of the magnetization
vectors of all of the atomic magnetic moments within its volume. Domains may
have different orientations with respect to each other depending on their net
energy content and the force vectors acting upon them. In some cases, domains
can be produced and moved about in a thin plate or layer in a reproducible way.
This is the case for a number of ferro-magnetic and ferrimagnetic materials.
Figure 9 shows a typical domain structure in a magnetic garnet at zero magnetic field with the random, worm-like domain patterns maintained, in the steady
state, by the inherent uniaxial magnetic anisotrophy of this material. If an
increasing magnetic field is applied perpendicular to the plane of the plate in
Fig. 9, then the unfavourably oriented domains (with respect to the applied field)
may be made to shrink and then finally to collapse into cylindrical domains which
look like "magnetic bubbles" if they are observed in polarized light under a
microscope; they can be displaced in the direction of an applied magnetic gradient
and their presence or absence at a certain position of the plate constitutes the
binary-coded information stored in the memory.

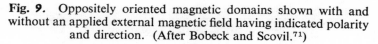

Fig. 9. Oppositely oriented magnetic domains shown with and without an applied external magnetic field having indicated polarity and direction. (After Bobeck and Scovil.[71])

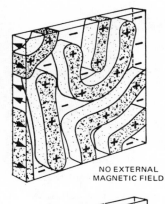

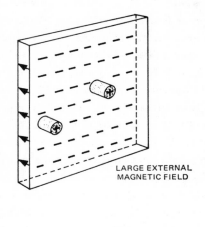

SOURCE: "Magnetic domain bubble memories" by J. L. Tomlinson and H. H. Weider. The Radio & Electronic Engineer. Vol. 45. No. 12. Dec. 1975. p. 727.

SEE ALSO: "Application of orthoferrites to domain-wall devices" by A. H. Bobeck, R. F. Fischer, A. J. Perneski, J. P. Remeika and L. G. Van Uitert. IEEE Trans. on Magnetics, MAG-5, pp. 544-54. 1969.

1969   "BUCKET-BRIGADE" DELAY CIRCUIT   F. L. J. Sangster
                                    K. Teer   (U. S. A.)

       The general principle is that the signal to be delayed is sampled and stored in a cascade of capacitors interconnected by switches operated at the same frequency as the signal sampler.

       As a new signal sample can evidently not be stored in a capacitor before the signal sample present is completely removed, only half the number of capacitors actually do store information at any moment, the others being empty.

       In the past only rather complicated circuitry has been proposed for this function, so that even in integrated form there was no chance for an inexpensive compact design. A much simpler solution presents itself when signal sample transfer is not established by a charge transfer in the direction of signal travel but in the opposite direction, by what is essentially a charge deficit transfer. This principle leads to a much simpler resistorless circuit suitable for realization in integrated--circuit form.

SOURCE: "Bucket-Brigade electronics - possibilities for delay, time-axis conversion, and scanning" by F. L. J. Sangster and K. Teer. IEEE Journal of Solid State Circuits. Vol. SC-4 No. 3. (June 1969) p. 131.

1969   MICROELECTRONICS (Bipolar)   Bell Laboratories   (U.S.A.)
        CDI (collector diffusion isolation)   Ferranti   (U.K.)

        In 1970 manufacturers began to investigate bipolar processes which
seemed to offer prospect of being competitive with m.o.s. For example there
was the c.d.i. process (collector diffusion isolation) developed first at Bell
Labs and then by Ferranti, the Isoplanar process of Fairchild, the Process
IV which was suggested at Plessey's research centre at Caswell, and the
Dutch Locos process developed by Philips. All of these were compatible with
circuits which could operate in excess of 1.5GHz and all of them had the
advantage of using less surface area than earlier processes. The c.d.i. system
for example, started with a slice of 10 to 20Ω. cm p-type silicon into which
n^+-layers were diffused. These were later to be the collectors of transistors
formed in a 1 Ω. cm p-type epitaxial layer put down on top of them. The n^+
diffusions were made through the epitaxial layer to make contact with the now
buried n^+ layers laid down at first. These not only acted as collector contacts
but isolated the area within. In this base area the n^+ emitter diffusion is made,
as well as any second emitter for a Shottky diode. After the oxide has been
deposited amd holes cut in it to gain access to the electrodes, silicon is grown
in the holes to the same level as the oxide, thus giving a flat surface.

SOURCE: "The semiconductor story" by K.J.Dean. Wireless World.
(April 1973) p.170.

SEE ALSO: "Collector diffusion isolated integrated circuits" by B.T.Murphy,
V.J.Glinski, P.A.Gary and R.A.Pederson. Proc.IEEE Vol.57.No.9.
(Sept.1969) p.1523.

1970   CHARGE COUPLED DEVICES   W.S.Boyle and G.E.Smith (U.S.A.)

        Storing charge in potential wells created at the surface of a semi-
conductor and moving the charge (representing information) over the surface
by moving the potential minima.

SOURCE: "Charge coupled devices - concepts, technologies and applications"
J.D.A.Benyon. The Radio & Electronic Engineer, Vol.45. No:11. November
1975. p.647.

SEE ALSO: W.S.Boyle and G.E.Smith "Charge coupled semiconductor devices"
Bell System Tech.J.Vol.49. (1970) p.583.

1970   X RAY LITHOGRAPHY FOR   E.Spiller   E.Castellani
        BUBBLE DEVICES   R.Feder   L.Romankiw   (U.S.A.)
                                     J.Topalian   M.Heritage

        X-rays have been used for several decades to obtain images of objects.
First proposals to use x-rays for the fabrication of microlectronic devices were
made in 1970. X-ray lithographic systems are very simple, they have a high
throughput because many wafers can be exposed simultaneously and they have a
resolution which is at least as good as that of electron beam systems. In 1972
Spears and Smith produced the first devices using proximity printing with x-rays

and demonstrated the high resolution capability of x-ray lithography. In particular, x-ray lithography lends itself to the fabrication of high resolution devices
requiring no alignment capability, such as magnetic bubble devices.

SOURCE: "X-ray lithography for bubble devices" by E. Spiller, R. Feder,
J. Topalian, E. Castellani, L. Romankiw and M. Heritage. Solid State Technology
(April 1976) p. 62.

SEE ALSO: "X-ray projection printing of electrical circuit patterns" by
R. Feder. IBM Report TR 22.1065, August 1970.

ALSO: "High resolution pattern replication using soft x-rays" by D. L. Spears
and H. I. Smith. Electron. Lett. Vol. 8. 102 (1972).

ALSO: "Evolution of bubble circuits processed by a single mask level" by
A. H. Bobeck, I. Danylchuck, F. C. Rossol and W. Strauss. IEEE Trans. on
Magnetics MAG-9, 474 (1973).

1971   LIQUID CRYSTAL STUDY OF OXIDE DEFECTS J. M. Keen (U. K.)

          One of the most convenient techniques was first reported by Keen.
It consists of introducing, between the oxidized silicon, and a tin oxide coated
glass slide, a thin film of negative nematic liquid crystal. On applying a
voltage across this "capacitor" structure, defects can be seen as highly
turbulent regions of liquid crystal. For plane electrodes without an oxide layer
the same turbulence is present everywhere. In the case of oxides containing
defect s, however, the turbulence is particularly violent making location of a
defect easy.

SOURCE: "Polarity dependent oxide defects located using liquid crystals" by
A. K. Zakzouk, W. Eccleston and R. A. Stuart. Solid State Electronics 1976.
Vol. 19. p. 133.

SEE ALSO: J. M. Keen Electron Lett. 7. 15. 432(1971)

1972   1024 BIT RANDOM ACCESS MEMORY   Intel (U. S. A.)

          The RAM - father of them all, the 1103 from Intel started the stam-
pede to semiconductor memories. It was the first time that more than 1000
bits of read/write memory could be supplied on a single semiconductor chip
in a low-cost MOS configuration.

SOURCE: "Special report - semiconductor RAM's land computer mainframe
jobs" by L. Altman.   Electronics (August 28, 1972) p. 64.

1972   AUTOMATIC CONTROL OF   W. Bardsley
        CRYSTAL GROWTH   G. W. Green
                                      C. H. Holliday   (U. K.)
                                      D. T. J. Hurle

          In this note, we describe a novel, alternative method of automatic
diameter control (or, more strictly, control of cross-sectional area, since the
crystals may be of non-circular section) for which certain advantages can be
claimed. Put simply, the method comprises "weighing" the growing crystal by

means of an industrial weighing cell from which the pull rod is hung. The method requires that there are no constraints to the vertical motion of the pull rod, and this is achieved by a gas bearing where the rod enters the growth chamber. Normally, some of the ambient gas escapes through the gas bearing, but for initial evacuation and flushing before growth, the rod is sealed by a constrictable rubber sleeve. The pull rod has a self-aligning bearing at its upper end to provide a connection to the weighing cell, and is rotated by a low friction pin and fork arrangement.

The electrical signal from the weighing cell is compared with a signal from a rectilinear potentiometer driven from the leadscrew nut and any difference is amplified and used to adjust the crucible heating power in that direction which minimuses the difference signal. The desired diameter is predetermined by setting electrically the magnitude of the potentiometer output voltage per unit distance of pull rod travel. The initial growth out from the diameter of the seed crystal to the final diameter has also been automatically controlled by introducing a non-linear element in series with the potentiometer output circuit.

SOURCE: "Automatic control of Czochralski crystal growth" by W. Bardsley, G. W. Green, C. H. Holliday and D. T. J. Hurle. Journal of Crystal Growth, Vol. 16 (1972) p. 277.

SEE ALSO: "Developments in the weighing method of automatic crystal pulling" by W. Bardsley, B. Cockayne, G. W. Green, D. T. J. Hurle, G. C. Joyce, J. M. Roslington, P. J. Tufton and H. C. Webber. Journal of Crystal Growth Vol. 24/25 (1974) p. 369.

1972   X-RAY SCANNER   E. M. I. (U. K.)

The skull surrounds the brain and provides a very good protection for this most delicate and vital organ, it also heavily attenuates diagnostic X-rays. The brain is a relatively homogeneous organ, when imaged by X-rays, which does not have much contrast to show up its structure. These two problems make imaging of the brain by conventional X-radiography of very limited diagnostic value. Contrast techniques can be used to improve the imaging but they do involve some risk to the patients and the need for hospitalisation. They are expensive.

In 1972, EMI Limited introduced computerised axial tomography to overcome these limitations. This radical new technique was developed at the Central Research Laboratories of EMI. Clinical trials rapidly showed that this was a major advance in diagnostic imaging.

In computerised axial tomography the patient is scanned by a tightly collimated narrow beam of X-rays. The transmitted beam is detected and converted to an electric signal after passing through the patient. Another detector is used in the reference mode to measure the primary X-ray beam.

The frame carrying the X-ray source and detectors traverses linearly across the patient, a large number of readings of X-ray intensity are taken and stored as it traverses, the gantry is then indexed round by a small angle and the process is repeated. This series of traverses and angular movements is repeated until a large matrix of data has been acquired.

The computer then uses this data to calculate the X-ray absorption co-efficient map of this cross section of the anatomy. This can then be display-ed as a brightness modulation map on a cathode ray tube or printed as a map of X-ray absorption numbers by a line printer. The computation cancels out the effects of absorption in other parts of the anatomy so that the problem of shadowing by the skull or bone structure is overcome.

SOURCE: "Section by Section" by Shelley Stuart. "Electronics Weekly" April 7th, 1976. p.16.

1972   MICROELECTRONICS (V-MOS technique)   T.J.Rodgers (U.S.A.)

Because of the work of a 27-year-old research engineer, American Microsystems Inc., in Santa Clara, California, is on the verge of committing itself in a big way to V-MOS - an n-channel metal-oxide semiconductor tech-nology that will compete with the new, faster and denser bipolar static designs and processes.

The engineer is T.J.Rodgers who, as a doctoral candidate in electrical engineering at Stanford University in nearby Palo Alto, invented the V-groove MOS process (Electronics, Sept 18. p.65.) His gaol was to push MOS technology to its limits so it would achieve bipolar speeds as well as high speed-power products and high packing densities in read-only and static random-access memories, random logic and microprocessor designs.

SOURCE: "Young EE's ideas to alter AMI's direction" Electronics. January 22,1976, p.14.

1974   16-BIT SINGLE CHIP MICROPROCESSOR   National (U.S.A.)

The semiconductor industry's first 16-bit, single-chip microprocessor is soon to be introduced by National Semiconductor Corp. Called PACE (for processing and control element), the device will handle 16-bit instructions and addresses, and either 16-bit or 8-bit data. It is being built with p-channel silicon-gate MOS technology because, the company says, p-MOS is a more predictable and better established technology than n-MOS and meets both of PACE's main requirements: 10-microsecond execution time for instructions, and enough density to fit the entire circuit on a single chip.

PACE requires only two power supplies, + 5 volts and -12 V, instead of the three required with n-channel fabrication.

SOURCE: "National to show 16-bit processor on single chip" Electronics. November 28th, 1974. p.35.

1975   MICROELECTRONICS - THIN FILMS   J.F.Burgess
        DIRECT BONDED COPPER PROCESS   C.A.Neugebauer   (U.S.A.)
                                         G.Flanagan & R.E.Moore.

In the direct copper to ceramic bonding process, bonding is accomplished by heating Al_2O_3 or BeO substrates in contact with the Cu foil. Foil thicknesses from 250 to 1 mil can be used. The gas atmosphere consists principally of inert gas such as argon or nitrogen with a small addition of oxygen, typically of the order of a few hundredths of a percent. The length of'time required for bonding is typically a few minutes.

The temperature for bonding is critically important. Bonding does not take place unless the temperature exceeds 1065°C, but it must be below 1083°C, which is the melting point of copper.

SOURCE: "Hybrid packages by the direct bonded copper process" by J. F. Burgess, C. A. Neugabauer, G. Flanagan and R. E. Moore. Solid State Technology. May 1975. p. 42.

1975   4096-BIT RANDOM ACCESS MEMORY   Fairchild (U.S.A.)

In a significant development, Fairchild Semiconductor has applied its oxide-isolated Isoplanar technology to an injection-logic configuration. The result: the industry's first 4,096-bit I^2L random-access memory. The part has a nominal access time of 100 nanoseconds, making it more than twice as fast as today's n-MOS 4-kilobit dynamic RAMs . The device will be ready for selective prototyping late this aummer.

SOURCE: "Fairchild develops first 4K RAM to use I^2L" Electronics. June 26, 1975. p. 25.

1975   MICROELECTRONICS (Integrated
        optical circuits)   F. K. Reinhart
                                         R. A. Logan   (U.S.A.)

For the first time, scientists have combined a laser with components such as modulators, filters, and lightguides in a single crystal microcircuit, just as multiple components are fabricated in an integrated electronic circuit.

The devices, integrated optical circuits measuring usually about 6 by 15 mils, operate within the structure of a semiconductor injection laser.

This type of circuit represents an alternative to hybrid integrated optics where components - often fabricated from different material systems - are interconnected on a base. By contrast, the new monolithic optical circuit "contains" many of the required components within the same single crystal.

Franz K. Reinhart and Ralph A. Logan of Bell Labs., Murray Hill, N. J., developed the new circuit.

SOURCE: "Integrated optical circuits: another step forward" Bell Labs. Record (Sept. 1975) p. 349.

1975   MICROELECTRONICS (Silicon anodisation)   R. Cook (ITT) (U.S.A.)

The discovery that silicon itself can be anodized opens an unexpected path to cheaper, denser, faster integrated circuits. The low-temperature process produces in one step the dielectric needed to isolate the active elements on a chip, thus adding the advantages of dielectric isolation to any semiconductor technology, whether bipolar or metal-oxide-semiconductor.

Direct silicon anodization was discovered quite by chance. An anodizing voltage was accidentally increased beyond the point required to anodize aluminum. The aluminium was destroyed, but the silicon substrate beneath the aluminium, when examined under a microscope, was seen to have been transformed into a porous dielectric layer. Further experiment revealed that the dielectric on the silicon surface could be tailored to almost any desired thickness simply by adjusting the anodizing process.

SOURCE: "Anodizing silicon is economical way to isolate IC elements" by B. Cook. Electronics. (November 13, 1975) p.109.

1976   MICROELECTRONICS (Versatile arrays)   Philips (Holland)

A simple variation of standard silicon-gate technology has produced extremely versatile arrays that make novel analog-to-digital converters, analog type displays, and light-pattern scanners. The arrays consist of devices similar to standard metal-oxide-semiconductor elements, except that a resistive electrode structure replaces the normal metal insulated gate. This structure permits a voltage gradient to be set up across the ends of the gate and then manipulated to control the transistors either singly or in groups.

SOURCE: "Resistive insulated gates produce novel a-d converters, light scanners" by M. V. Whelan, L. A. Daverveld and J. G. deGroot. Electronics. (March 18, 1976) p.111.

1976   MICROELECTRONICS
(16, 384 bit Random Access Memory)   Intel   (U.S.A.)

A triumph of semiconductor device technology, the 16, 384-bit random access memory has arrived. Its bit density is unprecedented and springs from an enhanced n-channel silicon-gate technique, in which a double level of poly-silicon conductors shrinks the memory cell to 450 micrometers square. That is less than half the cell size in the densest 4, 096-bit RAM.

SOURCE: "Enter the 16, 384-bit RAM" by J. E. Coe and W. G. Oldham Electronics (Feb.19th, 1976) p.114.

1976   COMPUTER (One board with
programmable 1/0)   Intel Corp. (U.S.A.)

A complete general-purpose computer subsystem that fits on a single printed-circuit board has been a major goal all through the steady evolution of LSI technology. Such a computer, consisting of a central-processing unit, read/write and read-only memories, and parallel and serial input/output - interface components, could satisfy most processing and control applications needed by original-equipment manufacturers. A single board computer could greatly extend the range of computer applications by providing a single solution to three problems that have often precluded the use of conventional computers.

The primary reason for use of a single assembly of LSI devices rather than a multiboard subsystem is economic. Extra board assemblies are costly in themselves and need related equipment, such as backplanes and housings, that also adds to cost.

Compactness and low power consumption are often prerequisites for

products.   Using LSI for all key computer functions reduces power consumption
and provides a higher functional density than conventional subsystem designs.
This new class of LSI devices - programmable input/output interface chips -
enables an 8-bit computer to be built as a subsystem on one printed-circuit
board.

SOURCE: "The 'super component': the one-board computer with programable
1/0" by R. Garrow, J. Johnson and M. Maerz.   Electronics. (February 5th, 1976)
p. 77.

Appendix 1   List of useful books on inventions

(N.B. - Some of the older books are now out of print and difficult to obtain.)

Abbott C.G. "Great Inventions"
Smithsonian Institution, Washington(1932)

Aitken W. "Who invented the telephone"
Blackie, London (1939)

Appleyard R. "Pioneers of electrical
communications" MacMillan, London(1930)

Barnouw E. "A history of broadcasting in
the United States" Oxford Univ. Press(1968)

Briggs Asa "The history of broadcasting in
the United Kingdom" Vols.1 & 2 Oxford
Univ. Press (1965)

Bryn E.W. "The progress of invention in the
nineteenth century" Munn. New York (1900)

Carr L.H.A. & Wood J.C. "Patents for
Engineers" Chapman & Hall, London (1959)

Carter E.F. "Dictionary of inventions &
discoverers" Fdk. Muller, London (1966)

Chase C.T. "A history of experimental
physics" Von Nostrand (1932)

Crowther J.G. "Discoveries & inventions
of the 20th Century" Routledge & Regan
Paul Ltd., London (1966)

Crowther J.G. "British Scientists of the
19th century" MacMillan, London.

Cressy E."Discoveries & inventions of
the twentieth century" George Routledge,
New York. E.P.Dutton (1914)

Darrow F.L. "Masters of science and
invention"Harcourt Brace, New York(1923)

Dibner Bern "Heralds of science as represented by 200 epochal books and
pamphlets selected from the Bundy Library"
Bundy Library, Norwalk, Conn. (1955)

Fahie J.J. "History of electric telegraphy
to the year 1837"F.N.Skoon, London (1884)

Fahie J.J. "History of wireless telegraphy"
Wm. Blackwood, Edinburgh & London (1899)

Fleming J.A. "Fifty years of electricity"
Wireless Press (1921)

Goldstine H.H. "The computer from Pascal
to von Neuman" Princeton Univ. Press(1972)

Hawks E. "Pioneers of wireless"
Methuen & Co. London (1927)

Jaffe Bernard "Men of science in America"
Simon & Schuster, New York (1944)

Jewkes J., Sawers D. & Stillerman R."The
sources of invention" MacMillan, London
(1958)

Larsen Egon "A history of invention" J.M.
Dent & Sons, London and Roy Publishers
New York (1971)

MacLaurin W.R. "Invention & innovation
in the radio industry" MacMillan, New York
(1949)

Moore C.K. & Spencer K.J. "Electronics-
a bibliographical guide" MacDonald,
London (1965)

Motteley "Bibliographical history of
electricity & magnetism" Griffin, London
(1922)

Pierce J.R. "The beginnings of satellite communications" San Francisco Press (1968)

Pledge H.T. "Science since 1500" H.M.S.O. (1946)

Proceedings of the Royal Society, London.

Rhodes F.L. "The beginnings of telephony" Harper, New York (1929)

Rider K.J. "The history of science and technology" Library Assoc.of London (1967)

Routledge R. "Discoveries & inventions of the nineteenth century" G.Routledge, London (1891)

Shiers G. "Bibliography of the history of electronics" The Scarecrow Press, Metuchen, N.J. (1972)

Singer C.J. "A short history of science to the nineteenth century"Clarendon Press (1941)

Tricker R.A.R. "Early electrodynamics" Pergamon Press, Oxford (1965)

Whetham W. "A history of science" Cambridge Univ.Press (1929)

Appendix 2   List of useful books on inventors

(N.B. Some of the older books are now out of print and difficult to obtain)

AMPERE "Andre-Marie Ampere and his English acquaintances" by K.R. and D.L. Gardiner Brit.Journal for the History of Science. Vol.2. (July 1965) p.235.

BAIRD "Baird of television - the life story of John Logie Baird" by R.F.Tiltman. Selley Service. London (1933)

BELL"Alexander Graham Bell; the man who contracted space" by Catherine D.Mackenzie Houghton Mifflin (1928)

"Bell, Alexander Graham Bell and the conquest of solitude" by R.V.Bruce Victor Gollanz (1973)

BERLINER 'Grevile Berliner, maker of the microphone" by F.W.Wile. Bobbs-Merrill, Indianapolis (1926)

"Biographical memoirs of Fellows of the Royal Society" Royal Society, London(Annual)

BRAUN "Ferdinand Braun: Leben und wirken des Erfinders der Brauchen Roehre, Nobel-preistraeger" by F.Kurylo Heinz Moos Verlag (1965)

CROOKES "The life of Sir William Crookes" by E.E.Fournier D'Albe. Fisher Unwin London (1923)

EDISON "My friend Edison" by H.Ford. Ernest Benn Limited.

"Edison" by M.Josephson. McGraw-Hill, New York (1959)

FAHIE "The life and work of John Joseph Fahie" by E.S.Whitehead. University Press of Liverpool (1939)

FARADAY "Faraday" by R.& R.Clark. Brit.Elec.& Allied Mnfrs. Assoc. (1931)

"Michael Faraday - his life and work" by S.P.Thompson. Cassel (1901)

"Faraday, Maxwell & Kelvin" by D.K.C.MacDonald. Doubleday. N.Y.(1964)

FESSENDEN "Fessenden - builder of tomorrow" by H.M.Fessenden. Coward-McCann (1940).

FITZGERALD "The scientific writings of the late George Francis Fitzgerald" by J. Lamor (Ed) Dublin Univ. Press (1902)

FLEMING "Memories of a scientific life" by Alexander Fleming. Marshall.

HEAVISIDE "Oliver Heaviside" by G. Lee. Longmans Green, London (1947).

HENRY "Joseph Henry - his life and work" by T. Coulson. Princeton Univ. Press U.S.A. (1950).

HERTZ "Gesammelte Werke" by P.E.A. Lenard Ambrosius Borth, Leipzig (1895) (Papers in three volumes) English translations - MacMillan, London.

LODGE "Oliver Lodge - past years, an autobiography" Scribner, New York (1932).

MARCONI "Marconi, the man and his wireless" by O. E. Dumlap, MacMillan (1937).

"Marconi, master of space" by B. L. Jacot & D. M. B. Collier, Hutchinson (1935)

"My father, Marconi" by D. P. Marconi. McGraw Hill. New York (1962).

"Marconi, pioneer of radio" by D. Coe, Julian Messner (1935).

MAXWELL "The life of James Clark Maxwell" by L. Campbell & W. Garnett MacMillan, London (1882).

"James Clark Maxwell F.R.S. 1831 - 1879" by R. L. Smith-Rose, Longmans Green, London (1948).

MORSE "The life of Samuel F.B. Morse inventor of the electro-magnetic recording telegraph" by S.A. Prime. D. Appleton, New York (1875).

REIS "Philipp Reis" Deutche Bundespost Archiv. fur Deutche Postgeschicht No. 1. (1963)

RONTGEN "Wilhelm Conrad Rontgen and the early history of the Rontgen rays" by O. Glasser. Charles C. Thomas, Springfield, Ill. (1934).

RUTHERFORD "Rutherford - being the life and letters of the Right Honourable Lord Rutherford" by A. S. Eve, Cambridge Univ. Press (1939).

TESLA "The inventions, researches and writings of Nikola Tesla" by T. C. Marti n The Electrical Engineer, New York (1894)

"Ten founding fathers of the electrical science" by B. Dibner. Bundy Library Publications, Norwalk, Conn. (1954) (GILBERT, GEURICKE, FRANKLIN, VOLTA, AMPERE OHM, GAUSS, FARADAY, HENRY and MAXWELL).

THOMSON J.J. "The life of Sir J.J. Thomson O.M. sometime Master of Trinity College, Cambridge" by Lord Rayleigh. Cambridge Univ. Press (1942)

"J.J. Thomson and the Cavendish Laboratory of his day" by G. P. Thomson, Nelson, London (1964).

THOMSON W. "The life of William Thomson" by S. P. Thompson. MacMillan London (1910).

INDEX   Inventions in alphabetical order.